Springer Series in Solid-State Sciences

Volume 199

The Springer Series in Solid-State Sciences features fundamental scientific books prepared by leading and up-and-coming researchers in the field. They strive to communicate, in a systematic and comprehensive way, the basic principles as well as new developments in theoretical and experimental solid-state science.

We welcome submissions for monographs or edited volumes from scholars across this broad domain. Topics of current interest include, but are not limited to:

- Semiconductors and superconductors
- Quantum phenomena
- Spin physics
- Topological insulators
- Multiferroics
- Nano-optics and nanophotonics
- Correlated electron systems and strongly correlated materials
- Vibrational and electronic properties of solids
- Spectroscopy and magnetic resonance

Artem V. Pronin

Linear Electrodynamic Response of Topological Semimetals

Experimental Results Versus Theoretical Predicitons

Springer

Artem V. Pronin
1st Physics Institute
University of Stuttgart
Stuttgart, Germany

ISSN 0171-1873 ISSN 2197-4179 (electronic)
Springer Series in Solid-State Sciences
ISBN 978-3-031-35636-0 ISBN 978-3-031-35637-7 (eBook)
https://doi.org/10.1007/978-3-031-35637-7

This Springer imprint is published by the registered company Springer Nature Switzerland AG
The registered company address is: Gewerbestrasse 11, 6330 Cham, Switzerland

Preface

This book is devoted to experimental optical studies of topological semimetals (TSMs) in comparison with theory predictions. Recently, the TSM phases came first in focus of theoretical studies [1–11] and then quickly found their material realizations [12–24]. Commonly, the term *topological semimetals* (or, sometimes, *topological metals* [10, 25, 26]) refers to the materials, whose three-dimensional electronic band structure exhibits band crossings or band touching points (the *nodes*) in the vicinity of the Fermi surface. The nodes are robust against small perturbations, i.e. topologically and/or symmetry protected. The *nodal semimetals*—another term utilized for TSMs—may contain not only point nodes, but also continuous lines [5, 10] and even surfaces [27, 28] of nodes in reciprocal space, making the variety of such three-dimensional topological phases extremely diverse [29]. The crossing or touching bands in TSMs can be characterized by non-zero topological indices, such as the Chern number. In a sense, TSMs are cousins of topological insulators (TIs) predicted and experimentally discovered a bit earlier [30–39].

The low-energy electronic dispersion relations for TSMs can be approximated by solutions of the Dirac equation or its modifications [40, 41]. This often makes the optical response of TSMs (due to the interband transitions) generally different from the response of "ordinary" metals and semiconductors and allows one to utilize the optical methods for probing the particular band structure of different TSMs. Experimental studies of the linear electrodynamic (optical) response of different TSMs and analysis of the obtained spectra are the scope of this work. A recent review on non-linear optical effects in TSMs can be found in [42].

As is implied from the above paragraph, the TSM phases are gapless. In real materials, however, small gaps of the order of tens of meV or less due to, i.e., spin-orbit coupling (SOC) can appear, while the band-structure calculations without SOC still predict gapless topological states. Studying such compounds can be of relevance for topological-material investigations, if the presence of these small gaps does not substantially affect the physical properties of the system. As argued below, this is the case, e.g., for ZrSiS—a material possessing a "gapped nodal line" [22].

Overall, this book contains (i) a basic introduction to the topological description of electronic bands, including those in TSMs; (ii) theory predictions for the linear electrodynamic response of TSMs; (iii) "experimental" chapters presenting the results and analysis of the optical studies on different TSMs (nodal-line, Weyl/Dirac, triple-point, and multifold semimetals); and (iv) a summary. A short description of the experimental setups used for the optical measurements is provided in the Appendix.

Each experimental chapter starts with a short introduction to the particular TSM type and to the representative compound(s) for this TSM type studied in this work. The material/sample characteristics and optical measurement parameters are provided in the relevant chapters for each material/sample. The obtained optical spectra and their interpretation are accompanied by discussions and conclusions for each material studied.

For frequency units, we use cm^{-1} and (milli)electronvolts. It is merely a tradition in the optical-conductivity community to utilize cm^{-1} for experimental results. The calculated optical-conductivity spectra are typically presented versus frequency in eV or meV. In the course of the book, we switch between these two types of units whenever we find this appropriate and we always discuss when we make such a change. Often, we provide graphs with two horizontal scales—in cm^{-1} and meV (or eV). For practical usage, 1 meV is roughly 8 cm^{-1} (higher accuracy is not necessary for the results presented here). The CGS system is used in the formulas and equations throughout the book.

Stuttgart, Germany
April 2023

Artem V. Pronin

Acknowledgements

Here, I would like to thank all the people who directly or indirectly contributed to this book and also those who provided me with a lot of support during the time I worked on it.

- First of all I would like to thank Martin Dressel—the director of the 1st Physics Institute (*1. Physikalisches Institut, PI1*) at the University of Stuttgart. All the experimental work presented in this book was done at this institute, and it would not be possible without the continuous support from Martin. He is a person who is able to encourage, and I am grateful to him for his encouragement during the years I worked at PI1.
- I am deeply indebted to the current and former postdocs and graduate students, who are or have been working at PI1 and who have contributed a lot to the measurements and data analysis: Ece Uykur, David Neubauer, Micha Schilling, Felix Hütt, Sascha Polatkan, Lucky Maulana, Anja Löhle, Weiwu Li, Richard Kemmler, Ievgen Voloshenko.

  I thank all of them, as well as Andreas Baumgartner, Tobias Biesner, Yuk Tai Chan, Sina Fella (née Zapf), Bruno Gompf, Eric Heintze, Ralph Hübner, Olga Iakutkina, Berina Klis, Helga Kumrić, Di Liu, Uwe Pracht, Samuel Pinnock, Andrej Pustogow, Laurin Rademacher, Seulki Roh, Roland Rösslhuber, Leonid Samoilenko, Marc Scheffler, Maxim Wenzel, and Run Yan, for many conversations on science and beyond, in the lab or over a glass of beer.
- I am very thankful to Gabi Untereiner for her technical assistance in preparing samples for optical experiments and to the institute's secretaries Agni Cienkowska-Schmidt and Olga Weber for countless occasions where administrative issues arose.
- I thank all the persons listed above as well as all other current and former members of PI1 for the great working atmosphere at the Institute.
- I appreciate very much the collaboration with Claudia Felser and the members of her group, particularly with Chandra Shekhar and Kaustuv Manna. They have provided the samples of excellent quality and of large dimensions, which is

essential for optical measurements. I am also indebted to Claudia for her help in establishing new collaborations.

- Other excellent samples discussed in this book have been provided by Leslie Schoop, Alex Nateprov, Ralph Hübner, Jürgen Nuss, Cuixiang Wang, and Youguo Shi. Additionally, I am grateful to Leslie for many fruitful discussions.
- I am indebted to the late Jules Carbotte for a collaboration on Cd_3As_2 and for his patient and very pedagogical explanations of different theoretical issues.
- I thank Alexander Yaresko, Zhi Li, Sascha Polatkan, and Ece Uykur for providing ab-initio calculations included in the book. I appreciate discussions with them, as well as with Anton Burkov, Mark Goerbig, Milan Orlita, Andreas Schnyder, Serguei Tchoumakov, Alexander Tsirlin, and Hongbin Zhang.
- I would also like to thank my good friends as well as my colleagues from different institutions all over the world—Katya Aksentova, Vasily Artemov, Alexander Boris, Geoffrey Chanda, Ivan Chapurin, Yan Dolomanov, Theo Fischer, Boris Gorshunov, Natasha Jones, Dima Kamenskyi, Ricardo Lobo, Andrei Malyshkin, Sergey Panin, Andrei Pimenov, Anya Pimenov, Ksenia Rabinovich, Igor Spektor, Marc Uhlarz, and Alexander Volkov—for the ongoing relationships with respect to many different subjects in physics and beyond.
- Finally, I would like to thank very much my mother Galina Ivanonva Pronina for her love and unconditional support to me in all areas of my life.

Contents

Chapter 1
Theoretical Background

The goal of this introductory chapter is two-fold. First, a pedagogical introduction to the electronic band topology is provided: we start from intuitively clear analogies between the three-dimensional geometrical objects and the electronic bands, introduce the concept of Berry phase and Chern number, and finally arrive at a simple description of such topologically nontrivial electronic phases as Chern insulators, topological insulators, and topological semimetals. This part follows in many aspects the book of Vanderbilt [43]. Second, we provide the theory expectations for the linear optical response of different types of TSMs, relevant for the experimental studies constituting the main part of this book. Finally, we briefly describe the theoretical expectations for the optical signatures of the chiral anomaly.

1.1 Berry Phases, Topological Indices, and Topological Electronic Bands

1.1.1 Relevance of Topology to Electronic Bands

Topology, as a branch of mathematics, studies the properties of geometric objects, which remain unchanged under continuous deformations such as stretching, twisting, crumpling, and bending. These unchanged properties can be expressed via topological indices or invariants—the integer numbers, which do not change under such deformations. A change of a topological index indicates a transition into a topologically different state and requires such discontinuous deformations as opening holes, closing holes, tearing, gluing, or passing through itself.

In basically any presentation on topology to a broad audience, one can find a picture, showing a mug and a doughnut, and a statement that the surfaces of these two objects are topologically indistinguishable. In Fig. 1.1, we provide a similar

A. V. Pronin, *Linear Electrodynamic Response of Topological Semimetals*,
Springer Series in Solid-State Sciences 199,
https://doi.org/10.1007/978-3-031-35637-7_1

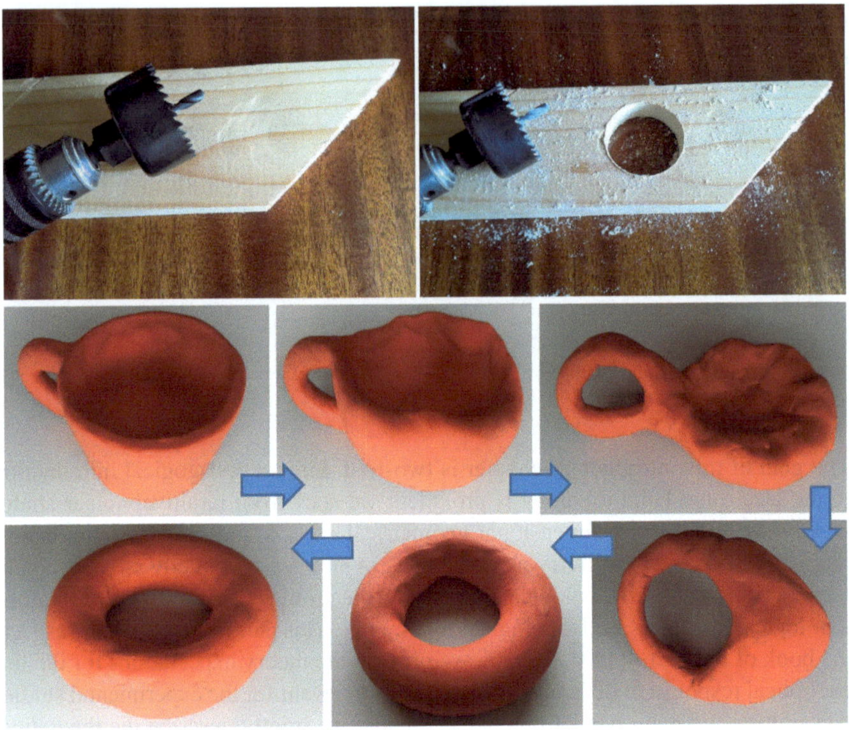

Fig. 1.1 Examples of a topological transition [drilling a through hole, upper row] and of the absence of such [transforming a Plasticine cup in a solid torus, lower rows]

example (transformation of a cup to a solid torus). We also show an example of a topological transition—making a through hole in an object.

Obviously, topology can deal with objects in many dimensions. Here, we will mostly consider two-dimensional (2D) surfaces in three-dimensional (3D) space, although we will also touch objects in other dimensions. To get an idea of a topological invariant, one can compare a sphere and a torus of revolution.[1] The former has no holes, while the latter has one through hole. The number of holes is called genus g and it is a topological invariant.

Importantly, such invariants can be expressed via integrals of the local (i.e., geometrical) quantities of the object under consideration, taken over the entire object. An example of this connection between the topological (global) and geometrical (local) properties is the Gauss-Bonnet theorem. It states that for a 2D orientable[2] surface with no edges (sphere and torus satisfy these requirements) the integral of

[1] Colloquially and also here, it is called just torus. For definitions of different types of tori, see https://en.wikipedia.org/wiki/Torus.

[2] Such surfaces have two sides. Möbius strip is not orientable, as it has only one side.

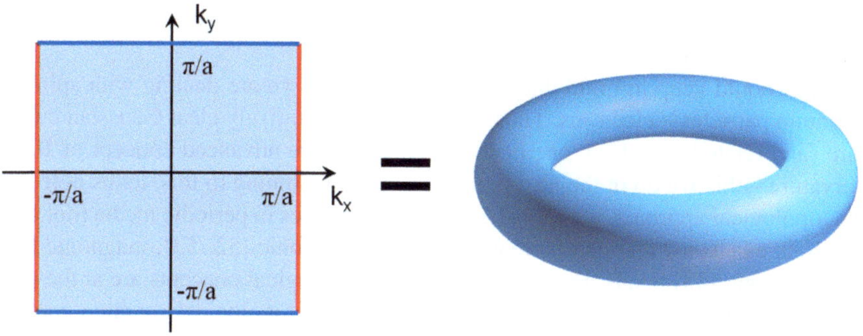

Fig. 1.2 A square 2D Brillouin zone is topologically equivalent to a torus

the Gaussian curvature K over the surface is 2π times an integer χ, called the Euler characteristic:

$$\oint_S K dS = 2\pi\chi. \tag{1.1}$$

The Gaussian curvature is defined as $K = 1/(r_1 r_2)$ with r_1 and r_2 being the maximal and the minimal radii of curvature for a given surface point. Note, that the radius of curvature possesses a sign, depending on what side of the orientable surface the center of curvature is situated. One can show that $\chi = 2$ for a sphere and 0 for a torus. Generally, $\chi = 2 - 2g$.

Topology enters physics in many different ways—the connection between topology and physics is very fundamental and not new. General introduction to the subject can be found in textbooks [44, 45]. Here, we concentrate on topological aspects of band-electron dynamics in crystalline solids. Materials with topologically non-trivial electronic bands became a central point of condensed matter physics in 2005, when topological insulators were theoretically discovered by Kane and Mele [30, 31]. By today, a large body of books [43, 46, 47] and review papers [23, 24, 48–51] is available on this subject.

In order to argue (on a very basic level) why topology can be relevant to electronic bands, let us consider a Brillouin zone (BZ) of a 2D square lattice with a lattice constant a (Fig. 1.2). The electronic states at the zone boundary at $k_x = -\pi/a$ are physically equivalent to the states at $k_x = \pi/a$ (the red vertical lines in Fig. 1.2). Hence, we can glue these states by wrapping the BZ around, to form a vertical cylinder. In turn, the states at $k_y = -\pi/a$ are equivalent to the states at $k_y = \pi/a$ (the blue horizonal lines). Thus, we can connect the ends of the cylinder and get a torus.

The dispersion relation of a given band, $E(\mathbf{k})$, maps the BZ onto the energy surface of this band. If such a band is completely (i.e., for any k) separated from any other neighboring band (obviously, there are many electronic bands in a solid), the shape of this band is topologically equivalent to a torus. However, if we have bands, which touch or cross each other, their shape will be topologically distinct from a torus, and for such bands $g \neq 1$ and $\chi \neq 0$. Thus, topological characteristics can be relevant for electronic bands.

One should keep in mind that in electronic bands we are dealing with spin-1/2 quantum particles—electrons. Thus, instead of the intuitively clear Gaussian curvature (and related g and χ), one needs to go to a more advanced concept of Berry curvature—this we do in the next paragraph. Before coming to this, let us note that the topological approach is applicable to different waves in periodic media (not only to Bloch electrons) and can be realized, e.g., in photonic [52, 53], magnonic [54] and phononic [55] bands. Furthermore, similar topological concepts are at the core of the modern theory of electric polarization [43, 56], topological quantum computations [47, 57], as well as many other topics of modern condensed matter physics [47, 58].

1.1.2 Berry Phase, Berry Connection, and Berry Curvature

In simple terms, a Berry phase (or a geometrical phase) is a phase acquired by a (quantum) system,[3] while it is carried around a path in its parameter space (often also called Hilbert space). The parameter space is the space of all possible states of the system. Each point of this space corresponds to a different state.

The definition of the Berry phase implies that the system must change its state to obtain a Berry phase, as otherwise there is no motion in parameter space. For example, the phase, acquired by a freely propagating plane wave in vacuum, is not a Berry phase, as the state of the wave (defined by its wave vector $\mathbf{k}$) doesn't change due to such motion. Instead, this phase is called dynamical phase. The phase, acquired by the same wave when it gets reflected from a surface is a Berry phase, because $\mathbf{k}$ is changed at reflection.

Let us consider an abstract quantum system, whose state is described by a complex vector (a wave function) $|u\rangle$. Let us imagine that the system changes its state in a discrete way from the initial state $|u_0\rangle$ via $|u_1\rangle$, $|u_2\rangle$, etc. to the final state $|u_N\rangle$. In the parameter space, we can visualize it as shown in Fig. 1.3a.

From quantum mechanics, we know that the phase difference between two quantum states, $|\psi_1\rangle$ and $|\psi_2\rangle$, is $\mathrm{Im}(\ln\langle\psi_1|\psi_2\rangle)$. (This is very easy to demonstrate for two scalar states, $|\psi_1\rangle = A_1 e^{i\varphi_1}$ and $|\psi_2\rangle = A_2 e^{i\varphi_2}$, remembering that A_1 and A_2 are real: $\ln\langle\psi_1|\psi_2\rangle = \ln(\psi_1^*\psi_2) = \ln(A_1A_2) + i(\varphi_2 - \varphi_1)$.) Hence, the phase acquired by our system while moving along the path shown in Fig. 1.3a, i.e., the Berry phase, is:

$$\phi = -\mathrm{Im}\{\sum_{n=1}^{N} \ln\langle u_{n-1}|u_n\rangle\}. \tag{1.2}$$

[3] Or by a complex vector in mathematical terms.

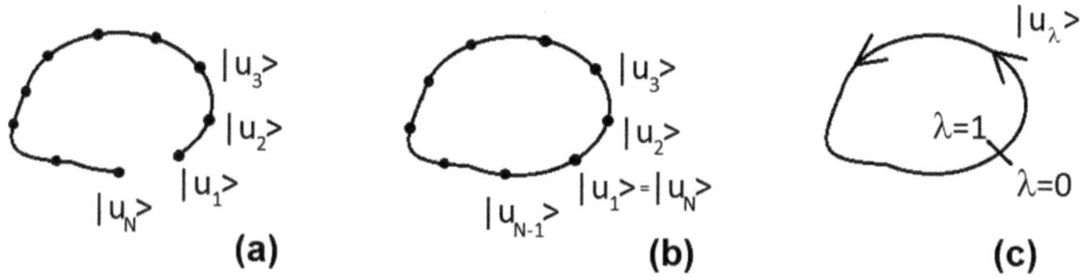

Fig. 1.3 Discrete (**a**, **b**) and continuous (**c**) paths of a quantum system in its parameter space; following [43]

The minus sign is a matter of convention here. Note that the Berry phase is only a function of the path, which the system traces in parameter space; it is independent of the rate, at which the path is traversed (at least for adiabatic evolution).

For our goal—to arrive at a topological invariant—we want to consider a closed path [Fig. 1.3b], i.e., the situation when $|u_0\rangle = |u_N\rangle$. Moreover, we take the continuous limit, where we can parameterize the states with a parameter λ changing continuously (with an infinitesimal step $d\lambda$) from 0 to 1 in such a way that $|u_{\lambda=0}\rangle = |u_{\lambda=1}\rangle$ [Fig. 1.3c]. The summation in (1.2) can be replaced now by integration along the path P in the parameter space:

$$\phi = -\mathrm{Im} \oint_P \ln\langle u_\lambda | u_{\lambda+d\lambda}\rangle. \tag{1.3}$$

The integrand in this equation can be simplified using the Taylor expansion and assuming $|u_\lambda\rangle$ to be smooth and differentiable:

$$\ln\langle u_\lambda | u_{\lambda+d\lambda}\rangle = \ln\langle u_\lambda | \left(|u_\lambda\rangle + d\lambda \frac{d|u_\lambda\rangle}{d\lambda} + \dots \right) = \\ \ln\left(1 + d\lambda\langle u_\lambda|\partial_\lambda u_\lambda\rangle + \dots\right) = d\lambda\langle u_\lambda|\partial_\lambda u_\lambda\rangle + \dots, \tag{1.4}$$

where ∂_λ is a shorthand for $d/d\lambda$. Thus, for ϕ we get:

$$\phi = -\mathrm{Im} \oint_P \langle u_\lambda|\partial_\lambda u_\lambda\rangle d\lambda. \tag{1.5}$$

Because $\langle u_\lambda|\partial_\lambda u_\lambda\rangle$ is purely imaginary ($2\mathrm{Re}\langle u_\lambda|\partial_\lambda u_\lambda\rangle = \langle u_\lambda|\partial_\lambda u_\lambda\rangle + \langle\partial_\lambda u_\lambda|u_\lambda\rangle = \partial_\lambda\langle u_\lambda|u_\lambda\rangle = 0$), we can also write:

$$\phi = \oint_P \langle u_\lambda|i\partial_\lambda u_\lambda\rangle d\lambda. \tag{1.6}$$

This expression is often used as a definition of the Berry phase. We arrived at it starting from an intuitively clear (1.2).

Equation (1.6) can be re-written as:

$$\phi = \oint_P A(\lambda) d\lambda \tag{1.7}$$

with

$$A(\lambda) = \langle u_\lambda | i \partial_\lambda u_\lambda \rangle. \tag{1.8}$$

$A(\lambda)$ is called Berry potential or Berry connection.

So far, we considered a 1D parameter space. We can generalize (1.5)–(1.8) to higher dimensions, where we get:

$$\phi = -\mathrm{Im} \oint_P \langle u_\lambda | \boldsymbol{\nabla}_\lambda u_\lambda \rangle d\boldsymbol{\lambda} = \oint_P \mathbf{A}(\boldsymbol{\lambda}) d\boldsymbol{\lambda}. \tag{1.9}$$

The Berry connection, $\mathbf{A}(\boldsymbol{\lambda}) = \langle u_\lambda | i \boldsymbol{\nabla}_\lambda u_\lambda \rangle$, is now a vector and $\boldsymbol{\nabla}_\lambda$ is a gradient in the parameter space.

Let us now recall that the phase acquired by a quantum particle with a charge e moving in the real space along a path l in magnetic field $\mathbf{B}$ is:

$$\phi = \frac{e}{c\hbar} \oint_l \mathcal{A}(\mathbf{r}) d\mathbf{r}, \tag{1.10}$$

where $\mathcal{A}$ is the vector-potential, $\mathbf{B} = \boldsymbol{\nabla} \times \mathcal{A}$. Based on the similarity of (1.9) and (1.10), one can say that the Berry connection in parameter space plays the same role as the vector-potential in real space. If we continue the analogy, we can introduce an analogue of magnetic field acting in parameter space:

$$\boldsymbol{\Omega} = \boldsymbol{\nabla}_\lambda \times \mathbf{A}. \tag{1.11}$$

This is the Berry curvature. Using this definition and Stokes' theorem, (1.9) can be transformed to:

$$\phi = \int_S \boldsymbol{\Omega}(\boldsymbol{\lambda}) d\boldsymbol{\lambda}, \tag{1.12}$$

where the surface S is bounded by the closed path P. (See [43] for more details on application of Stokes' theorem here.) Hence, Berry curvature can be interpreted as Berry phase per unit area in parameter space.

Similarly to the Gauss-Bonnet theorem (1.1), there is a connection between an integral of Berry curvature and a topological invariant: according to the Chern theorem the integral of $\boldsymbol{\Omega}$ over a closed 2D surface is 2π times an integer C called Chern number:

$$\oint_S \boldsymbol{\Omega} d\mathbf{S} = 2\pi C. \tag{1.13}$$

As we will see below, Chern numbers play a very important role in distinguishing the topological states.

It can be shown that the Berry curvature and, consequently, the Chern number are gauge invariant, while the Berry connection is not. This continues the analogy with the magnetic field: $\mathbf{B}$ is gauge independent, $\mathcal{A}$ is not. The Berry phase for a closed path is gauge invariant modulo 2π (i.e., the Berry phase after a gauge transformation is the original Berry phase plus $2\pi m$, where m is an integer). The Berry phase for an open path is not gauge invariant.

1.1.3 Time-Dependent Phase

As known from elementary physics textbooks, the dynamical phase, that is the phase acquired as a function of time without any change in the system's (eigen)state $|n\rangle$, is $\gamma(t) = E_n t/\hbar$, where E_n is the energy of this state. Then the system's wave function evolves with time as $|\psi(t)\rangle = e^{-i\gamma(t)}|n\rangle$. A plane Bloch wave with a fixed $\mathbf{k}$ obviously acquires such a phase while propagating in a crystal.

So far, we discussed the Berry phase as a property of a quantum system traveling along a path $P(\lambda)$ in parameter space. The parameter λ generally depends on time: $\lambda = \lambda(t)$ and the system's Hamiltonian is time dependent (leading to the time-dependent eigenstates $|n(t)\rangle$). The time evolution of such a system is described by a time-dependent Schrödinger equation and the solutions of this equation for a slow (adiabatic) change of λ can be represented as [43]:

$$|\psi(t)\rangle = e^{-i\gamma(t)} e^{\phi(\lambda(t))} |n(t)\rangle, \tag{1.14}$$

where $\phi(\lambda(t))$ is the time-dependent Berry phase (which is effectively an open-path Berry phase):

$$\phi(\lambda(t)) = \int_{\lambda(0)}^{\lambda(t)} A_n(\lambda) d\lambda, \tag{1.15}$$

and $A_n(\lambda) = \langle n(\lambda)| i\partial_\lambda n(\lambda)\rangle$. Hence, the total phase acquired by the system as a function of time is basically (up to a sign) a sum of the dynamical and geometrical (Berry) phases.

1.1.4 "Berryology" in Electronic Bands

Electrons in a crystalline solid are described by the Bloch waves:

$$\psi_{n\mathbf{k}}(\mathbf{r}) = e^{i\mathbf{k}\mathbf{r}} u_{n\mathbf{k}}(\mathbf{r}). \tag{1.16}$$

Here n labels the bands, $\mathbf{r}$ and $\mathbf{k}$ are the election position and momentum, and $u_{n\mathbf{k}}(\mathbf{r})$ are cell-periodic functions, as usually.

It can be argued that in order to define the Berry phase for Bloch electrons, one needs to utilize the cell-periodic functions $u_{n\mathbf{k}}$, rather than the full Bloch functions $\psi_{n\mathbf{k}}$, because the cell-periodic functions encode the details of the band structure, while the factor $e^{i\mathbf{kr}}$ arises from translation symmetry, intrinsic to any periodic lattice. For an electron in the n-th band, which is completely separated from all neighboring bands, the Berry phase and the Berry connection are defined as (cf. 1.9):

$$\phi = \oint \mathbf{A}_n(\mathbf{k})d\mathbf{k}, \ \mathbf{A}_n(\mathbf{k}) = \langle u_{n\mathbf{k}}|i\nabla_{\mathbf{k}}u_{n\mathbf{k}}\rangle. \tag{1.17}$$

Then the Berry curvature is:

$$\boldsymbol{\Omega}_n(\mathbf{k}) = \nabla_{\mathbf{k}} \times \mathbf{A}_n(\mathbf{k}). \tag{1.18}$$

Here, $\nabla_{\mathbf{k}}$ is a gradient in the reciprocal space. Note, that the Berry curvature depends on $\mathbf{k}$ and hence is a "local" property of the band (similarly to the Gaussian curvature, which is also defined locally).

For a 2D electronic band, the Berry connection and curvature can be re-written as:

$$\begin{aligned} A_{n\mu}(\mathbf{k}) =&\langle u_{n\mathbf{k}}|i\partial_\mu u_{n\mathbf{k}}\rangle, \\ &\Omega_{n,\mu\nu}(\mathbf{k}) = \partial_\mu A_{n\nu}(\mathbf{k}) - \partial_\nu A_{n\mu}(\mathbf{k}), \end{aligned} \tag{1.19}$$

with $\{\mu, \nu\} = \{k_x, k_y\}$ and $\partial_\mu = \partial/\partial k_\mu$.

One can further show that for such a 2D band, the Chern theorem (1.13) is modified to:

$$2\pi C_n = \int_{BZ} \Omega_{n,k_xk_y} dk_x dk_y, \tag{1.20}$$

where the integration is taken over the entire BZ. Equation (1.20) connects a local band property—the Berry curvature—to a topological index, related to the entire band—the Chern number. In the 3D case, the Chern numbers can be computed for 2D cuts of the 3D BZ, see [43] for details.

Usually, the bands with $C_n = 0$ are referred to as "topologically trivial" or simply "trivial". The bands with $C_n \neq 0$ are then "topologically non-trivial" or just "topological".

The Berry curvature $\Omega_n(\mathbf{k})$ for the n-th band is a gauge-invariant uniquely defined function of $\mathbf{k}$. Based on the above definitions of $\Omega_n(\mathbf{k})$, one can show that:

1. In crystals with inversion symmetry $\Omega_n(\mathbf{k}) = \Omega_n(-\mathbf{k})$;
2. In crystals with time-reversal (TR) symmetry $\Omega_n(\mathbf{k}) = -\Omega_n(-\mathbf{k})$, hence all quantities involving an integral of $\Omega_n(\mathbf{k})$ over the entire BZ (for example, the Chern number) vanish;

Fig. 1.4 Example of a crystal structure without inversion symmetry: the Weyl semimetal TaP (TaAs, NbAs, and NbP possess the same structure)

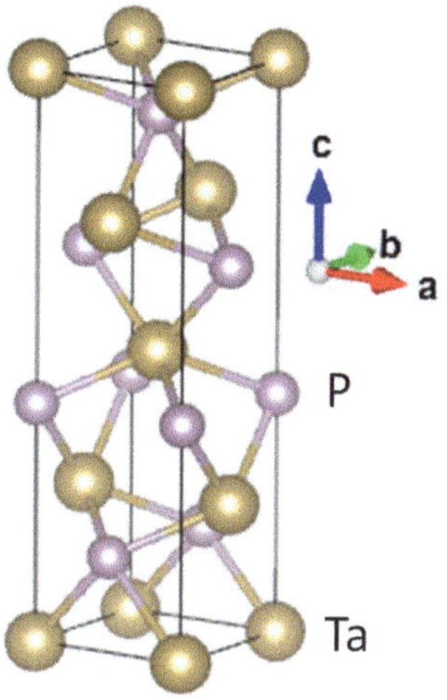

3. As follows form the points above, for crystals with inversion and TR symmetries, $\Omega_n(\mathbf{k}) = 0$ for any $\mathbf{k}$.

Let us note that the majority of materials current solid-state physics is dealing with (such as Si, Ge, or superconducting MgB_2) do have inversion symmetry. An example of the crystal structure for a material possessing no inversion symmetry is shown in Fig. 1.4. Let us also recall that TR symmetry implies the absence of magnetization (and no external magnetic field).

1.1.5 Wave Packets and Anomalous Velocity

When we measure the electronic properties of materials, for example, the transport properties, we do not deal with the single Bloch waves, but with wave packets. The quasiclassical approximation is often relevant for such measurements. As the wave packets propagate through the material, they disperse (i.e., become broader) in real and in reciprocal spaces. The evolution of the components of the wave packets can be described by (1.14).

From classical and quantum physics, we know that the propagation of the wave packets in real space is described by their group velocity, v_g. It can be shown relatively easily that the presence of Berry phase necessarily leads to an additional "group" velocity, which is commonly called anomalous velocity, v_a. A simple derivation of this can be found in [59].

For the total velocity of the wave packet, one gets (here, $\mathbf{r}$ is the position vector of the wave-packet center):

$$\dot{\mathbf{r}} \equiv v_{tot} = v_g + v_a = \frac{1}{\hbar}\frac{d}{d\mathbf{k}}E(\mathbf{k}) + \dot{\mathbf{k}} \times \boldsymbol{\Omega}(\mathbf{k}). \tag{1.21}$$

This velocity is an observable: one can determine it, e.g., from the current density, $\mathbf{j} = en\dot{\mathbf{r}}$ (n is the carrier concentration). Thus, Berry phase (or Berry curvature, which is Berry phase per unit area in parameter space, see 1.12) is directly related to physical observables.

The quasiclassical description of wave packets leads to the well-known equation of motion:

$$\dot{\mathbf{k}} = \frac{e}{\hbar}\mathcal{E} + \frac{e}{\hbar}\dot{\mathbf{r}} \times \mathbf{B}, \tag{1.22}$$

which directly follows from the standard ansatz for the Lorentz force, $\mathbf{F} = e\mathcal{E} + e\dot{\mathbf{r}} \times \mathbf{B}$ ($\mathcal{E}$ is the external electric field). Equations (1.21) and (1.22) constitute a new set of equations describing the propagation of an electron wave packet in a solid with non-zero Berry curvature. One can notice that the presence of the second term in the right-hand side of (1.21) makes the two equations look similar (without this term, (1.21) is reduced to the standard expression for v_g in an electronic band). It is also apparent now that $\boldsymbol{\Omega}$ plays the role of magnetic field in reciprocal space.

1.1.6 Chern-Insulator State and Integer Quantum Hall Effects

The Chern insulator is likely the simplest, from the theory point of view, electronic topological state. To demonstrate the idea of how this state can emerge, let us calculate the Hall (i.e., transverse) conductivity using the equations of motion (1.21) and (1.22) and implying that only external electric field is applied. Let us recall, that in the case of the ordinary Hall effect it is the magnetic field, which causes Hall conductivity to appear. What we intend to demonstrate here is that for a material with non-zero Berry curvature, Hall current is possible even without $\mathbf{B}$.

In the simple formula for electric current mentioned above ($\mathbf{j} = en\dot{\mathbf{r}}$), it is assumed that all electrons have the same velocity. Obviously, band electrons possess different velocities and we should modify this formula to:

$$\mathbf{j} = e\sum_{\mathbf{k}} \dot{\mathbf{r}}(f_{\mathbf{k}} + g_{\mathbf{k}}), \tag{1.23}$$

where $f_{\mathbf{k}}$ and $g_{\mathbf{k}}$ are the equilibrium Fermi-Dirac distribution and the deviation from it due to the applied $\mathcal{E}$ field, respectively (see, e.g., [60] for details).

The conductivity tensor is defined via $j_l = \sigma_{lm}\mathcal{E}_m$ (l and m are Cartesian indices). It has diagonal (ohmic conductivity) and off-diagonal (Hall conductivity) elements. For our purposes, we only need to consider the off-diagonal ones. Thus, we will not be interested in $g_{\mathbf{k}}$, as it is important only for the ohmic conductivity, where the equilibrium contribution sums up to zero (we introduced $g_{\mathbf{k}}$ in (1.23) just for completeness).

Combining (1.21), (1.22), and (1.23) and keeping in mind that a vector product of two vectors is perpendicular to both of them, we arrive at:

$$\mathbf{j}_{\text{Hall}} = \frac{e^2}{\hbar}\mathcal{E} \times \sum_{\mathbf{k}} \mathbf{\Omega}(\mathbf{k}) f_{\mathbf{k}}. \tag{1.24}$$

Assuming the electric field is directed along x and the Hall current along y, for the Hall conductivity we obtain:

$$\sigma_{\text{Hall}} \equiv \sigma_{yx} = \frac{e^2}{\hbar}\sum_{\mathbf{k}} \mathbf{\Omega}(\mathbf{k}) f_{\mathbf{k}}. \tag{1.25}$$

For $T = 0$, $f_{\mathbf{k}} = 1$ for occupied states and 0 otherwise, thus we get:

$$\sigma_{yx} = \frac{e^2}{\hbar}\sum_{\mathbf{k}}^{occupied} \mathbf{\Omega}(\mathbf{k}). \tag{1.26}$$

What we obtained here is the Hall current in the case of intrinsic anomalous Hall effect (AHE) [61]. AHE is the Hall effect in zero external magnetic field. This effect is present, for example, in magnetized ferromagnetic materials. Recall that the bands in ferromagnetic materials are spin-polarized, the TR symmetry is broken, and hence the sum of $\mathbf{\Omega}(\mathbf{k})$ for all $\mathbf{k}$ (in the relevant bands) can be different from zero, cf. the last paragraph in Sect. 1.1.4.

Imagine now that we have all electronic bands either completely occupied or empty, i.e., we have an insulator. In this case, the summation in (1.26) (for any of the occupied bands) goes over the entire BZ and the Hall conductivity is:

$$\sigma_{yx} = \frac{e^2}{\hbar}\sum_{\mathbf{k}}^{BZ} \mathbf{\Omega}(\mathbf{k}) = \frac{e^2}{\hbar}\int_{BZ} \mathbf{\Omega}(\mathbf{k}) d\mathbf{k}. \tag{1.27}$$

In 2D case, it can be written as:

$$\sigma_{yx} = \frac{e^2}{\hbar}\int_{BZ} \Omega(k_x, k_y) dk_x dk_y. \tag{1.28}$$

Comparing (1.28) and (1.20), one immediately recognizes that the Hall conductivity of a 2D insulator is quantized:

$$\sigma_{yx} = \frac{e^2}{h} C \tag{1.29}$$

(C in the Chern number, an integer). If $C \neq 0$, there is a non-zero *quantized* AHE conductivity and one speaks about the quantum anomalous Hall effect (QAHE). The quantization unit is e^2/h, one half of the conductance quantum G_0 ($G_0 = 2e^2/h$, the factor 2 in this definition comes from the electron spin degeneracy).

In fact, what matters for a non-zero QAHE is the sum of the Chern numbers for all occupied bands. For many materials $C \equiv 0$ for any band and/or the sum of all Chern numbers is zero. The 2D insulating materials with $C \neq 0$ in zero magnetic field are called Chern insulators. A state similar to this can be realized in thin films of topological insulators (TIs) doped with magnetic ions, see, e.g., [62] and Fig. 1.5. Non-magnetic TIs will be discussed below. Here, we just note that magnetic TIs are very rare. So far, there are only a few materials, where this state was suggested.

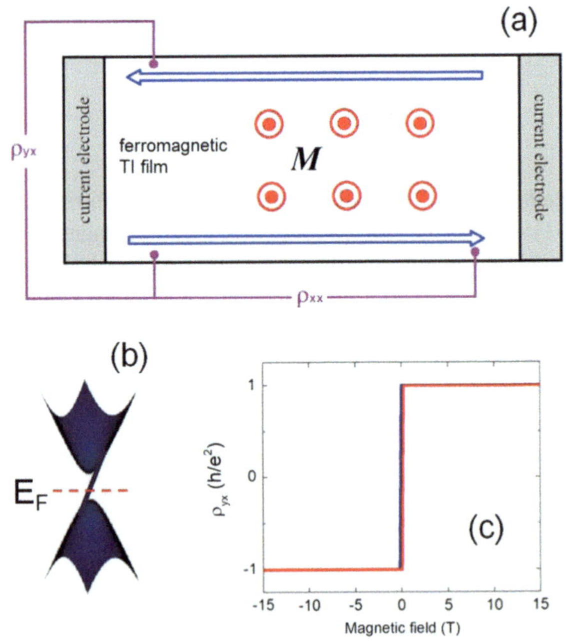

Fig. 1.5 Anomalous quantum Hall effect and Chern insulators: relevant experiments. Panel **a**: Schematic diagram of the experimental setup used in [62]. Thin film of a TI doped with magnetic ions ($Cr_{0.15}(Bi_{0.1}Sb_{0.9})_{1.85}Te_3$) mimics a Chern-insulator state. The Cr spins are oriented, producing a non-zero magnetization **M** and broken TR symmetry. The chiral edge current is shown as long horizontal arrows. Panel **b**: A sketch of the Chern-insulator band structure near the Fermi level E_F showing the gapped bulk bands and the chiral edge state crossing E_F. Panel **c**: Schematic representation of Hall resistance ρ_{yx} obtained in [62] for $Cr_{0.15}(Bi_{0.1}Sb_{0.9})_{1.85}Te_3$ versus the external magnetic field. The quantization is perfect in units of h/e^2. Note that there is a small hysteresis (red and blue curves correspond to different directions of the field sweeps) and that $\rho_{yx} \neq 0$ in zero field, while the longitudinal ohmic conductivity (not shown) vanishes

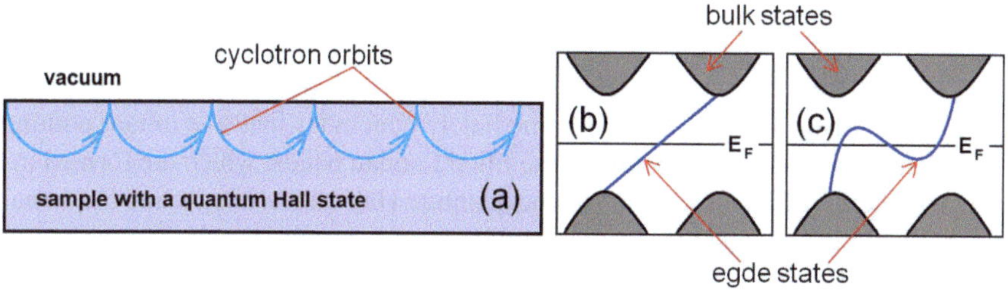

Fig. 1.6 Anomalous quantum Hall effect and Chern insulators: electron trajectories and band-structure diagrams. Panel **a**: Skipping electron orbits near the sample edges. The picture is valid for QAHE as well as for "ordinary", e.g., not anomalous, quantum Hall effect. Panels **b** and **c**: 1D projections of the band structures possible in Chern insulators. Each crossing point of the edge state and E_F corresponds to a conduction channel

Oppositely, the non-magnetic TI state is currently confirmed in many different materials.

The presence of Hall currents in a 2D bulk insulator can be explained by edge currents, which appear at the interface between samples with a quantum Hall state and vacuum. The currents may be seen as skipping orbits that electrons follow when their cyclotron orbits bounce off the sample edges, see Fig. 1.6a. Let us quote Kane here [96]: "... the electronic states responsible for this motion are chiral in the sense that they propagate in one direction only along the edge. These states are insensitive to disorder because there are no states available for backscattering—a fact that underlies the perfectly quantized electronic transport in the quantum Hall effect".

A schematic diagram of the band structure for a Chern insulator is demonstrated in Fig. 1.6b: the bulk is fully gapped (insulating) and there is a single conduction edge channel, which intersects the Fermi level E_F with a positive group velocity, $dE/dk > 0$, and can be defined as, for example, a right moving chiral edge mode. Let us note, that the dispersion relation of the edge states $E(k)$ can in principle be modified by changing the Hamiltonian near the surface (e.g., by adding some impurities). For instance, $E(k)$ can develop a kink, as shown in Fig. 1.6c. This new $E(k)$ intersects E_F three times—twice with a positive group velocity (right movers with $N_R = 2$ conduction channels) and once with a negative group velocity (left movers, $N_L = 1$). The difference between the number of right moving and left moving modes, however, is an integer topological invariant characterizing the interface and cannot change, as it is determined by the topological structure of the bulk states [96]. This is summarized by the bulk-boundary correspondence: $N_R - N_L = \Delta C$, where ΔC is the difference in the Chern number across the interface ($N_R - N_L = C$ for a sample-vacuum interface).

The integer quantum Hall effect in an external magnetic field can be explained in a similar way—via the skipping orbits and edge currents. The Hall conductivity in this case is also quantized and is defined via the TKNN number n (after Thouless, Kohmoto, Nightingale and den Nijs):

$$\sigma_{yx} = \frac{e^2}{h} n. \tag{1.30}$$

From comparison with (1.29), one immediately concludes that n is in fact nothing, but the sum of the Chern numbers of the filled Landau bands, which are formed in a magnetic field.[4] More information on the quantum Hall effect (integer and fractional) can be found in [64] and references therein.

1.1.7 *Topological Insulators*

In the previous section, we have shown that 2D insulators with broken TR symmetry can be topologically classified according to their Chern number C. This is known as a $\mathbb{Z}$ classification, because $C \in \mathbb{Z}$. For nonmagnetic insulators $C = 0$ (TR symmetry is preserved in this case), but, as we argue below, such insulators can also be topologically classified, i.e., some of such insulators are topologically different from others.

Let us consider a system, in which the spin-up and spin-down subsystems are decoupled—the projection of spin on the z axes (s_z) is a good quantum number. In practice, this can be realized in the systems with SOC. However, SOC should not be too strong to mix the spin projections. In this case, we can define Chern numbers $C_\uparrow$ and $C_\downarrow$ separately for each spin subsystem. The total Chern number is $C_{\text{tot}} = C_\uparrow + C_\downarrow$. Thus, we can introduce a new topological classification, in which each system (i.e., each material) is labeled by a pair of integers $(C_\uparrow; C_\downarrow)$. This can be called a $\mathbb{Z} \times \mathbb{Z}$ classification. Since we consider the case of a TR-symmetric system, $C_{\text{tot}} = 0$ and, consequently, $C_\uparrow = -C_\downarrow$. Hence, such systems can also be characterized by a single topological index, say $C_\uparrow$, and we arrive at a $\mathbb{Z}$ classification for nonmagnetic (TR-symmetric) insulators. Below we will show that for such systems $C_\uparrow$ can only take two values, 0 and 1.

Let us now consider a TR-symmetric 2D insulator with $C_\uparrow = 1$. If an electric field $\mathcal{E}$ is applied along the $+x$ direction, spin-up electrons will move in the $+y$ direction, producing a current $+(e^2/h)\mathcal{E}$, because $C_\uparrow = +1$ is assumed and (1.29) must be fulfilled. Spin-down electrons will in turn move in the opposite direction ($C_\downarrow = -1$ for such electrons) and provide an exact cancellation of the charge current. One can describe this situation by saying that there is an induced spin current, which flows in the $+y$ direction. In other words, we have a *spin* Hall effect. Moreover, this spin current is quantized and one can speak about a quantum spin Hall effect (QSHE) [32, 33]. A 2D material demonstrating this effect is called a 2D topological insulator (2D TI).

A schematic band structure of a 2D TI can be constructed in the following way. For spin-up electrons, we basically have a band structure of a Chern insulator with

---

[4] Each Landau level can be considered as a flat narrow band; recall that the quantum Hall effect is observed in a 2D electron gas.

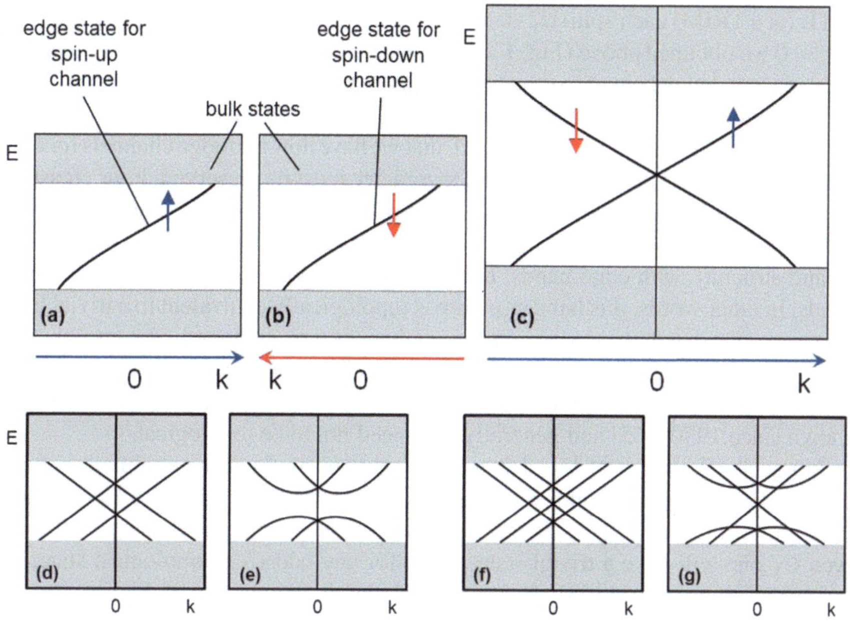

Fig. 1.7 Schematic band structures of 2D topological insulators. Panels **a** and **b**: band structures for spin-up and spin-down electrons in a 2D TI with the Chern numbers (1; −1), see text. Note opposite k directions in the two panels. Panel **c**: the total (combined) band stricture. Panels **d** and **e**: bulk and edge bands for a TR-symmetric material with the Chern numbers (2; −2). The crossings occurring not at a TRIM ($k = 0$ in our case) must gap, as show in panel (**e**). This leads to a trivial band structure. Panels **f** and **g**: same bands in the case of Chern numbers (3; −3). Gaps are opened (panel (**g**)), but the crossing at $k = 0$ survives, leading to a topological state. The Fermi level is assumed to lie in the bulk gap for all panels

insulating bulk and a single edge channel crossing E_F, see Fig. 1.7a. The TR symmetry reverses both, spin and momentum. Thus, for spin-down electrons the band structure should look identical to the band structure for spin-up electrons, but with the reversed directions of s_z and k, as shown in Fig. 1.7b. The total band structure is then a superposition of both diagrams, as shown in Fig. 1.7c. The patterns are now mirror-symmetric about $k = 0$, because of the TR symmetry, and the edge bands cross at this point.

So far, we assumed that s_z is a good quantum number. In reality, the spin-mixing terms, related to the s_x and s_y spin components, are always present in the SOC Hamiltonian. Such mixing typically leads to avoided crossings, i.e., to gapping of the crossing bands. This, however, doesn't happen at certain points of the BZ—the time-reversal invariant momenta (TRIM). TR is supposed to have no influence on a TRIM. As TR transforms $\mathbf{k}$ into $-\mathbf{k}$, at a TRIM $\mathbf{k}$ is such that it is identical to $-\mathbf{k}$ modulo a reciprocal lattice vector. The center of the BZ is obviously a TRIM. In 1D, another TRIM is $k = \pi/a$ (a is the lattice constant). There are four different TRIM points per BZ in 2D and eight in 3D. The Kramers theorem states that in the presence of TR (at a TRIM) each spin-1/2 state is at least twice degenerate. Thus, the crossing at $k = 0$ we obtained above (Fig. 1.7c) is robust, as it appears at a TRIM. In real TIs, the crossings indeed appear at the center of BZ [48, 49].

Now, what happens, if our 2D TI has $(C_\uparrow; C_\downarrow) = (2; -2)$? A schematic band stature for this case is shown in Fig. 1.7d. Recall, that we have now two edge channels for each spin projection and that the overall TR symmetry must be preserved. Four crossings of the edge states are now apparent. Only two of them, however, are situated at a TRIM. The other two must develop gaps, as shown in panel (e). Thus, we arrive to a band structure with edge bands, but these bands do not connect the gapped bulk bands. In other words, this band structure is topologically equivalent to a trivial band structure—the band structures can be adiabatically transformed one into another without such "violent" events as a gap closure. The presence of the surface states doesn't automatically make the band structure non-trivial. In fact, surface states are known since 1930s [65] and generally they need not to be topological.

In panels (f) and (g) of Fig. 1.7, we show that using the same arguments—band crossings become gapped, unless they happen at a TRIM—one arrives to a topological state for $(C_\uparrow; C_\downarrow) = (3; -3)$. In the same fashion, one can demonstrate that for any even $C_\uparrow$ one will have a trivial state, while for any odd $C_\uparrow$ a topological state. As D. Vanderbilt points out [43]: "A topological classification exists if there are two or more classes such that no member of one class can be converted into a member of another by following some adiabatic path without gap closure". Hence, we can say that for TR-symmetric insulators we have two topologically distinct classes: with odd and even Chern numbers. Indeed, one cannot topologically distinguish the $C_\uparrow = 0$ state from the $C_\uparrow = 2, 4, 6$, etc. states—all these states are fully gapped. Similarly, the $C_\uparrow = 1$ state is not distinguishable from the states with $C_\uparrow = 3, 5, 7$, etc., as all these states possess two chiral edge channels, which connect the bulk valence and conduction bands and cross at $k = 0$. Thus, it is sufficient to have only two possible Chern numbers: 0 and 1. This is the famous $\mathbb{Z}_2$ topological classification: $C_\uparrow \in \mathbb{Z}_2 \equiv \{0, 1\}$.

A TI state can also be realized in 3D: the bulk of these materials is insulating, while the surfaces conduct. The surface bands cross at a TRIM point and form Dirac cones, similar to the ones observed in graphene (the crossing points are often called nodes or Dirac points). The electronic excitations in such bands possess linear dispersion and can be referred as (massless) Dirac fermions. The 3D TIs can be subdivided into strong and weak TIs and are characterized by four Chern numbers, each can be 0 or 1. In strong TIs, all surfaces are conducting, while in weak TIs only some of them are. A weak 3D TI can be imagined as a stack of 2D TIs. Such a TI has four conducting and two insulating surfaces. A pedagogical introduction to 3D TIs can be found in [43].

As mentioned above, the 2D and 3D TI states are realized in many different materials. The most well-known examples include HgTe quantum wells for the 2D states and Bi_2Te_3, Bi_2Se_3, and related compounds for 3D. More information on experimental realizations of TI states can be found in [48, 49].

1.1.8 Topological Semimetals

There are materials, in which the topological properties are also related to 3D bulk bands, not only to the surface states [23, 24, 47, 66–69]. These are topological semimetals—the focus of this book. In such materials, the bulk electronic bands cross or touch each other and the integrals of the Berry curvature over certain 2D cuts[5] of the 3D BZ produce non-zero Chern numbers. The band crossing/touching points are called nodes, as in the case of TIs. We note here that there are also 3D materials possessing such nodes, but with the Chern numbers being 0 for any 2D cut of the BZ. Many properties of such materials are similar to the properties of topological semimetals and hence in the literature they may also be referred to as topological semimetals. Perhaps, a more general term—nodal semimetals—is more appropriate to describe the materials with nodes in their electronic structure, independently of the possible Chern numbers. We also note that the nodes may form lines (the nodal-line semimetals) and even surfaces in reciprocal space. A few examples of the diverse variety of different 3D nodal electronic phases are shown in Fig. 1.8. Currently, bulk nodal phases are well confirmed by the means of ARPES, scanning tunneling spectroscopy, and other experimental methods in many different materials [24].

The most studied cases of nodal semimetals are Dirac and Weyl semimetals. In both cases, linear electronic bands cross at a number of points in reciprocal space (see upper row in Fig. 1.8, where only a single crossing point is shown for each example). Note that the crossing bands form cones in 4D—three **k** directions and energy. The difference between the Dirac and Weyl semimetals is the (effective) spin degeneracy of the crossing bands: in Dirac semimetals the bands are spin degenerate, while in Weyl semimetals they are not. The electronic excitations in Weyl or Dirac bands (the Weyl or Dirac fermions, respectively) are effectively massless and possess linear dispersion.

For realization of a Weyl-semimetal state, either inversion or time-reversal symmetry should be broken, see [26] and references therein. TaP is an example of such a state (Fig. 1.4). In a Weyl semimetal, the band structure always contains *pairs* of the nodes (also called Weyl points), which are either sinks or sources of Berry curvature. This leads to non-zero Chern numbers, calculated over the 2D surfaces around the nodes. Half the Weyl points has positive (+1), the other half (−1) negative Chern number. In Weyl semimetals, the Chern number can be associated with electron chirality, implying that the (effective) spin can be either parallel or antiparallel to momentum (counted from the node) for the electrons in the Weyl bands near the nodes.

The presence of the Weyl nodes leads to the surface Fermi arcs, see Fig. 1.9b–d. To understand what a Fermi arc is, let us recall that in a 3D material at $T = 0$ the highest occupied states form a surface—the Fermi surface. In 2D, this will be the Fermi contour. Important is that the Fermi surfaces and Fermi contours are closed objects in reciprocal space. Fermi arcs exist in the surface BZ, i.e., they are 2D objects similar to Fermi contours, but Fermi arcs are not closed, as shown in Fig. 1.9d.

---

[5] Because of the BZ periodicity, such cuts can be viewed as 2D closed surfaces.

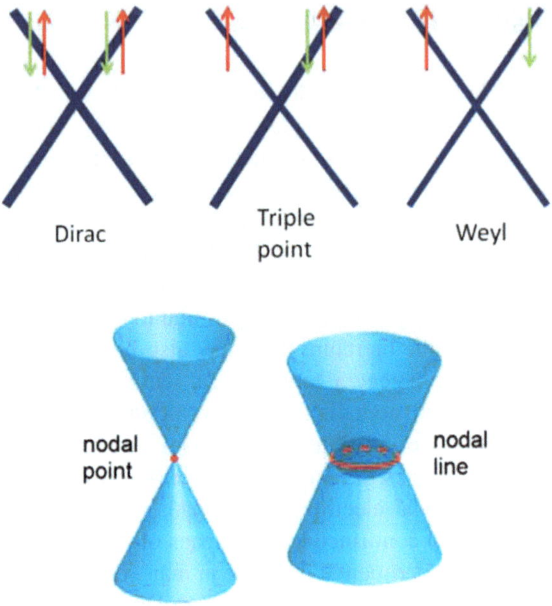

Fig. 1.8 Schematic examples of the band structures in a few different types of nodal semimetals (energy and momentum are plotted along the vertical and horizontal axes, respectively). The upper row shows one-dimensional band-structure cuts for Dirac, triple-point, and Weyl semimetals. For all cases, only a single node is shown. The difference is the spin degeneracy of the crossing bands involved. The bottom panel compares a "nodal-point" (e.g., Dirac or Weyl) semimetal and a nodal-line semimetal in a 2D cut of reciprocal space. In real materials the nodal lines may possess different shapes

The occurrence of Fermi arcs in a Weyl semimetal can be understood in the following way. In Fig. 1.9b, a 3D BZ of a Weyl semimetal is shown. The Weyl nodes are marked with stars. Following Vanderbilt [43], let us consider a cylinder like that shown in this panel enclosing one Weyl point. Its base is a circle in the two "horizontal" momentum directions parameterized by λ going from 0 to 1, and it extends fully across the BZ in the third direction. Since the top and bottom of the BZ are identical, the cylinder is in fact a closed surface (a torus), and carries a nonzero Chern number given by the chirality of the enclosed Weyl point. Panel (c) shows a sketch of the surface band structure plotted versus λ. Because of the nonzero Chern number, a surface state is topologically required to cross the bulk gap, as shown. The crossing of this state with E_F is indicated by the heavy dot in this panel and also in the view of the surface BZ in panel (d), where the loop is again indicated. Since we can repeat this construction for cylinders of different radii, and since we are required to find a crossing with E_F at any radius, the dot in panel (d) must actually belong to a Fermi arc extending out of the projected location of the Weyl point as shown. Hence, each Weyl point acts as a source, from which a Fermi arc is emitted. But the arc has to terminate, and can only do so at the surface projection of another Weyl

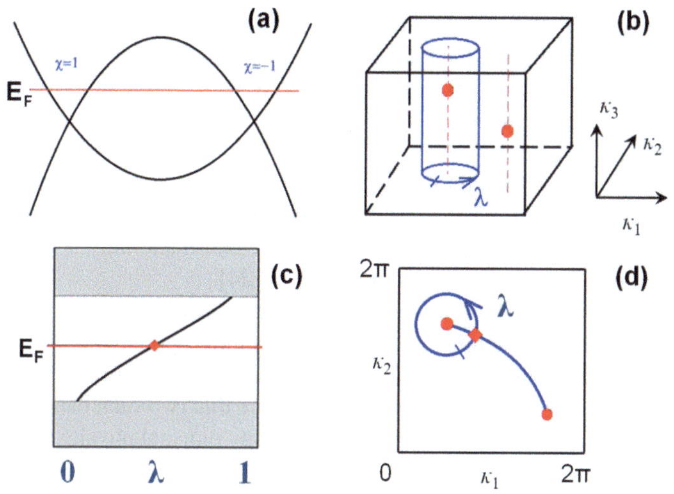

Fig. 1.9 Weyl semimetals: bulk band structure and surface Fermi arcs. Panel **a**: two Weyl nodes with opposite chiralities. Energy axis is vertical, momentum axis is horizontal. Note that the crossing bands are non-spin-degenerate. Panel **b**: 3D BZ of a Weyl semimetal with two Weyl points shown by bold red dots. The blue cylinder shows a surface enclosing one of the Weyl points, following a path in the (κ_1; κ_2) projected plane parameterized by λ running from 0 to 1. Panel **c**: surface band structure on the (001) surface plotted vs. λ. Existence of such an up-crossing surface state implies a Chern number of +1 on the cylinder (cf. Fig. 1.6b). The bold red diamond shows a crossing of the surface state with Fermi energy. Panel **d**: surface BZ with the Fermi arc, which connects projected positions of the Weyl points. If an orthorhombic unit cell is considered for simplicity, $\kappa_1 = k_x/a_1$, $\kappa_2 = k_y/a_2$, and $\kappa_2 = k_z/a_3$ ($\mathbf{a}_1$, $\mathbf{a}_2$, and $\mathbf{a}_3$ are the primitive lattice vectors). The figure follows [43]

point with the opposite chirality, as also shown in panel (d). The Fermi arcs are yet another example of a bulk-boundary correspondence. On the opposite-surface BZ, there is another Fermi arc, which connects the projections of the Weyl points, giving rise to an "ordinary" closed electron (or hole) pocket for the Weyl-semimetal slab as a whole. If more than two pairs of Weyl points are present in the 3D BZ, then multiple Fermi arcs will appear in the surface band structure. In fact, the majority of real Weyl semimetals possess multiple pairs of nodes related by crystal symmetries.

The separation of Weyl points in reciprocal space is the source of another peculiar feature of these materials—the so-called chiral anomaly [70–72], i.e., a violation of the charge conservation law for electrons of a given chirality χ near a Weyl point.[6] Note that the conservation law for the total number of electrons is obviously preserved. For electrons of the chirality $\chi = \pm 1$, the continuity equation in the presence of external electric and magnetic fields takes the form:

$$\frac{\partial \rho_\chi}{\partial t} + \nabla \mathbf{j}_\chi = \chi \, \frac{e^3}{ch^2} \, \mathcal{E} \cdot \mathbf{B}, \tag{1.31}$$

[6] Not to be confused with the Euler characteristic, for which the same letter is used.

here ρ_χ and $\mathbf{j}_\chi$ are the chiral-electron density and current, respectively. Recall, that in ordinary materials and for the total electron density in Weyl semimetals, the right-hand part of this equation is 0. Thus, the charge density at a single Weyl point is not conserved in the presence of $\mathcal{E}$ and $\mathbf{B}$ fields. This is the essence of the chiral anomaly. The missing (extra) charge at a given Weyl point is compensated at another Weyl point of the opposite chirality to ensure the overall conservation of charge in the system. The application of (parallel) $\mathcal{E}$ and $\mathbf{B}$ fields can be used to drive charges between the Weyl points of opposite chirality. The chiral anomaly is a source of the most unusual properties of Weyl semimetals [23, 24].

Let us also note that Dirac nodes may split in external magnetic fields, leading to a field-induced Weyl semimetal state and the chiral anomaly. Also, the number of non-degenerate bands crossing at a node is generally not limited to two. In the so-called multifold or multi-Weyl semimetals, more that two such bands cross at one point, leading to higher Chern numbers and to many unusual physical properties, as discussed, e.g., in [11].

To conclude this section, we point out that there is no symmetry that distinguishes two different topological states from each other. There is no order parameter, which is zero in one of such states and becomes nonzero in another one. It is the topological indices that distinguish different topological states. Above, we attempted to go from the very basics of topology via introducing the general topological properties of band electrons towards the topological semimetals (TSMs, as abbreviated earlier), which are in focus of this book. The idea was to make this path as short as possible, but logically clear. Next, we describe the theoretical expectations for the electrodynamic (optical) response of different TSMs.

1.2 Electrodynamics of Topological Semimetals

1.2.1 Complex Optical Conductivity and Dielectric Function

This book is concerned with the linear local response [73–75]. The experimental findings are typically expressed in terms of the frequency-dependent complex optical conductivity, $\sigma(\omega) = \sigma_1(\omega) + i\sigma_2(\omega)$, or the complex permittivity (the dielectric function), $\varepsilon(\omega) = \varepsilon_1(\omega) + i\varepsilon_2(\omega)$, which are related to each other via:

$$\varepsilon(\omega) = 1 + i\frac{4\pi}{\omega}\sigma(\omega). \tag{1.32}$$

Here, ω is the angular frequency. Hereafter, we will also use the ordinary frequency, $\nu = \omega/2\pi$, and often express it in cm^{-1} or in energy units (meV and eV). The magnetic permeability is normally considered to be unity, as the compounds studied here are either non-magnetic or have been investigated in paramagnetic phase only. We also note that the frequencies of our interest (>50 cm^{-1}) are above the typical frequencies of magnetic modes in zero external **B** field, cf. [76–78]. Overall, only the electro-dipole transitions are relevant for our study. The electro-dipole absorption mechanisms associated with various excitations in TSMs (as well as, in any other material) give rise to additive contributions to the spectra of $\sigma_1(\omega)$. This allows probing the exotic extinctions discussed above (such as Weyl fermions) by optical spectroscopy.

As $\sigma(\omega)$ and $\varepsilon(\omega)$ are generally second rank tensors and because the studied compounds often possess symmetry lower than cubic, thereafter we indicate what component(s) of these tensors have been studied for each investigated compound. Typically, we dealt with a tetragonal crystal symmetry and probed the in-plane response.

1.2.2 Electronic Band Dispersion and Optical Conductivity

In this section, we briefly recap theoretical predictions for the frequency behavior of optical conductivity in the major types of TSMs. In this book, we study nodal-line, Weyl, Dirac, triple-point, and multifold semimetals. Their schematic band structures are shown in Figs. 1.8 and 1.10. Generally, optical conductivity can be computed using the Kubo formula [60, 75, 79]. More details on computations for specific cases can be found in vast available literature, especially in the works of Carbotte and coauthors [80–89].

In the case of electron-hole symmetric d-dimensional bands with $E(k) \propto |k|^z$, the real part of the interband optical conductivity is supposed to follow a power-law frequency dependence [90, 91],

$$\sigma_1^{\text{IB}}(\omega) \propto \omega^{(d-2)/z}. \tag{1.33}$$

For Dirac and Weyl semimetals, $d = 3$, the bands are conical ($z = 1$), and (1.33) can be more specifically rewritten as

$$\sigma_1^{\text{IB}}(\omega) = \frac{e^2 N_W}{12h} \frac{\omega}{v_F}, \tag{1.34}$$

where N_W is the number of Weyl nodes (for a single Dirac node, $N_W = 2$), v_F is the Fermi velocity, $h = 2\pi\hbar$ is the Planck constant, and all Weyl/Dirac bands are considered to be identical (up to a spin degree of freedom) with their nodes situated at the chemical potential ($\mu = 0$). If the node position is not at the chemical potential ($\mu \neq 0$), transitions for the energies below 2μ are Pauli-blocked, and (1.34) is modified to

$$\sigma_1^{\text{IB}}(\omega) = \frac{e^2 N_W}{12h} \frac{\omega}{v_F} \theta\left\{\hbar\omega - 2\mu\right\}, \tag{1.35}$$

where $\theta\{x\}$ is the Heaviside step function and any carrier scattering is ignored. In this case, an intraband contribution to conductivity will also be present in the spectra.

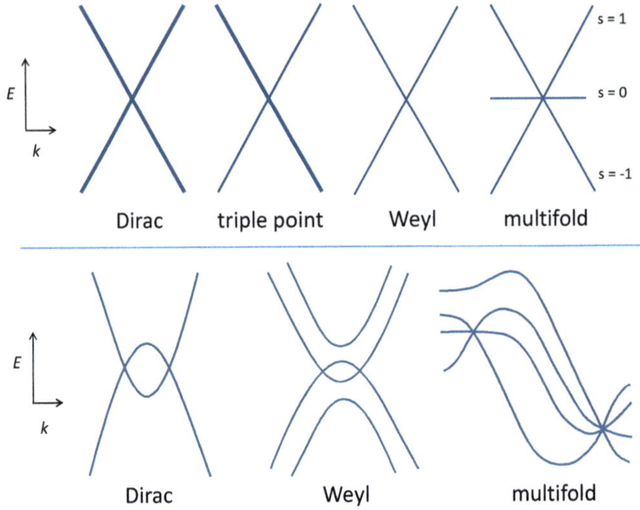

Fig. 1.10 Model band structures for different TSMs. Upper row: schematic band dispersions near the nodes in Dirac, triple-point, Weyl, and multifold semimetals (from left to right). The multifold-semimetal dispersion is presented for a spin-1 threefold-fermion case. Bottom row: more realistic (but still schematic) band structures, related to the compounds from this book—Dirac (Cd_3As_2), Weyl (the TaAs family), and multifold semimetals (RhSi and PdGa). For the latter case, the electronic bands are shown without SOC being included. For the realistic band structures of triple-point semimetals, please refer to the corresponding chapter below. The band degeneracy is encoded as the line thickness—thin lines represent non-degenerate bands, while thick lines (doubly) degenerate bands

For finite electron scattering, the Heaviside function can be replaced, for example, by

$$\frac{1}{2} + \frac{1}{\pi} \arctan \frac{\omega - 2\mu/\hbar}{\gamma} \tag{1.36}$$

with γ representing an appropriate scattering rate and the intraband conductivity can be approximated by a standard Drude ansatz [75], see Fig. 1.11 for a few examples.

In [81], it was shown that tilting the conical bands (relevant, e.g., for type-II Weyl semimetals [9, 92]) affects the linear behavior of optical conductivity: $\sigma_1^{\text{IB}}(\omega)$ remains (quasi)linear, but experiences slope changes at certain frequency points, which positions are related to μ and to the tilt angle.

For generalizations of Weyl bands with higher Chern numbers [6, 11, 93–96], the shape of $\sigma_1^{\text{IB}}(\omega)$ depends on the band dispersion relations. In the multifold semimetals, where a few linear (rotationally symmetric) bands with generally different slopes cross at a given point of the BZ [6, 11], the optical conductivity is linear in frequency (up to the steps, related to the Pauli-blocked transitions) [97]. For more complicated band structures, such as touching bands with a linear dispersion in one direction and parabolic dispersions in the remaining two [93, 94], $\sigma_1^{\text{IB}}(\omega)$ is expected to be anisotropic [98], in accord with (1.33). Additionally, if the nodes are situated at

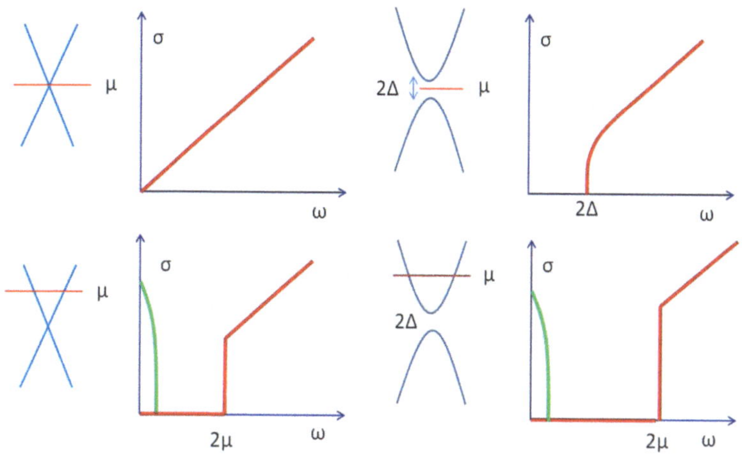

Fig. 1.11 Sketch of the optical conductivity (real part) for a 3D electronic system. Left-hand panels: Weyl or Dirac semimetals with crossing linear bands. Right-hand panels: parabolic bands. The chemical potential μ is situated at energies, where the density of states is either zero (top panels) or not (bottom panels). Note the appearance of the intraband (Drude-like) mode shown in green for the latter case. The interband transitions according to (1.33), (1.34), (1.35) are provided in red

different energies, as appears, e.g., in real multifold semimetals, $\sigma_1^{\mathrm{IB}}(\omega)$ changes its frequency run at different energy scales. The important point is that the total interband $\sigma_1(\omega)$ can often be decomposed into contributions from the nodes of each kind, simplifying interpretation of experimental spectra.

A particularly interesting case is the nodal-line semimetals [5], where the presence of a continues line of nodes effectively reduces the dimensionality of the crossing electronic bands to $d = 2$. This reduced dimensionality leads to a frequency-independent $\sigma_1^{\mathrm{IB}}(\omega)$ according to (1.33). Earlier, such "flat" optical conductivity has been predicted and experimentally observed in graphene and graphite [99–101] with a universal conductance value per one graphene sheet, $\pi e^2/(2h)$. In nodal-line semimetals, no universal sheet conductance is expected; instead $\sigma_1^{\mathrm{IB}}(\omega)$ is related to the length of the nodal line k_0 in the BZ [84, 85, 102]. For a circular nodal line, one has:

$$\sigma_1^{\mathrm{IB}}(\omega) = \frac{e^2 k_0}{16\hbar}. \tag{1.37}$$

It is assumed here that the plane of the nodal circle is perpendicular to the electric-field component of the probing radiation and that there is no particle-hole asymmetry. For $\mu \neq 0$, a Pauli edge (1.35 and 1.36) occurs in the conductivity spectra.

In the case of the so-called dispersive nodal line, i.e., when the energy position of the nodal line is momentum dependent, the interband optical conductivity restores its linearity in frequency, as was shown in [103]. In this case, $\sigma_1^{\mathrm{IB}}(\omega)$ is proportional to the slope of the nodal line in reciprocal space.

1.3 Chiral Anomaly as Seen in Optics

1.3.1 Chiral Anomaly in Steady Fields

As follows from the Weyl-bands description provided in Sect. 1.1.8, the chiral electronic excitations exist in the vicinity of Weyl nodes, half of which possessing the positive (effective spin and momentum are parallel), another half negative (antiparallel) chiralities. Simultaneous application of magnetic and electric fields is supposed to pump the electrons towards the nodes of a given chirality at the expense of the opposite chirality, the pumping effect being proportional to the scalar product of $\mathcal{E}$ and **B**, cf. (1.31).

Many theoretical investigations have been devoted to this solid-state realization of chiral anomaly, see [23] for a review. The experimental results reported on the chiral anomaly are mostly based on dc-transport measurements—the negative longitudinal magnetoresistance and the planar Hall effect, see, e.g., [104–106]. It is however known that dc transport may suffer from other ("parasitic") effects, such as current jetting appearing in anisotropic and/or inhomogeneous samples [107, 108]. Thus, special measurement procedures have to be applied to distinguish the chiral-anomaly induced transport effects from possible spurious features [109]. Reports on detecting the chiral anomaly by other experimental methods, such as optics, are therefore of high importance, but remain relatively rare [110–112]. The advantage of optical measurements is that the current-jetting effects can be circumvented, as for experimental determination of optical conductivity no contacts are required.

Theoretically, it has been shown [80, 113] that the low-frequency (terahertz or far-infrared) optical measurements may provide a way to observe the chiral anomaly in Weyl semimetals, see Fig. 1.12, where calculations of the real part of optical conductivity for a realistic Weyl semimetal are shown for the $\mathcal{E} \cdot \mathbf{B} = 0$ and $\mathcal{E} \cdot \mathbf{B} \neq 0$ cases. Two important features appear upon application of the fields: (i) splitting of the onset of the interband transitions at the frequency, corresponding to 2μ (μ is the chemical potential for $\mathcal{E} \cdot \mathbf{B} = 0$, cf. 1.35) and (ii) the Drude weight increases (note that the electronic scattering rate remains the same independently on the fields applied). The splitting of the onset of interband transitions is a direct consequence of the charge pumping between the chiral bands, see the inset in Fig. 1.12. The increased intraband conductivity, in turn, directly reflects the negative dc magnetoresistance and is related to the increased total number of free carriers, which provide an additional contribution to the plasma frequency, as argued by Son and Spivak [114]. This increased free-carrier response can be detected as an increase of the Drude spectra weight (Fig. 1.12).

The field-induced changes in the intra- and interband responses can in principle be detected experimentally. Such experiments are however very challenging. The major problem is the large dc conductivity of Weyl semimetals (typically, $10^4 - 10^5 \Omega^{-1}$ cm^{-1}). This prevents usage of metallic electrodes for creating strong enough electric fields within the Weyl-semimetal samples. Using isolating linings in turn leads to strong electric polarization inside such samples. The induced polarization

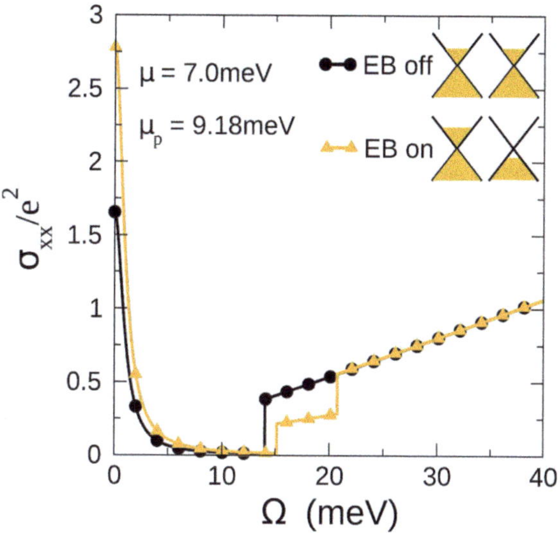

Fig. 1.12 Optical conductivity of a realistic doped Weyl semimetal (μ is not at the Weyl nodes) at T = 0. The black curve shows the results for $\mathcal{E} \cdot \mathbf{B} = 0$, while the dark yellow curve represents the case when $\mathcal{E} \cdot \mathbf{B} \neq 0$. The charge pumping leads to occurrence of two unequal chemical potentials for the electrons with different chiralities, as depicted in the inset. The shifts of the chemical potentials in the chiral bands can be found via: $\mu_\chi^3 = \mu^3 \pm \mu_p^3$, where μ_χ stands for the chemical potential in the Weyl band of a given chirality and μ_p determines the chemical-potential shift due to the electron pumping, $\mu_p^3 \propto \mathcal{E} \cdot \mathbf{B}$. The magnetic-field-induced features due to the Landau levels are assumed to be negligible, i.e., **B** is low. Reprinted with permission from [80]. Copyright (2014) by the American Physical Society

compensates (or significantly weakens) the external field $\mathcal{E}$. Still, a few attempts to detect the chiral anomaly in this way have been undertaken. One of them is described below in Sect. 3.5.1.

1.3.2 Dynamic Chiral Anomaly

An alternative way to detect the presence of the chiral bands by optical means is to employ the so-called *dynamic* chiral anomaly. In this case, the steady electric field $\mathcal{E}$ is replaced by the ac electric field of the electromagnetic waves $\mathbf{e}(t)$, a constant external **B** field being still applied.

In Fig. 1.13, the idea of the dynamic chiral anomaly is schematically explained for an undoped Weyl semimetal. In zero external magnetic field, the chemical potential for the cones of both chiralities is at the Weyl nodes. When parallel **B** and **e** fields are applied, the chemical potentials of the two cones start to change dynamically: they oscillate following the amplitude of the $\mathbf{e}(t)$ field. These oscillations are in anti-phase

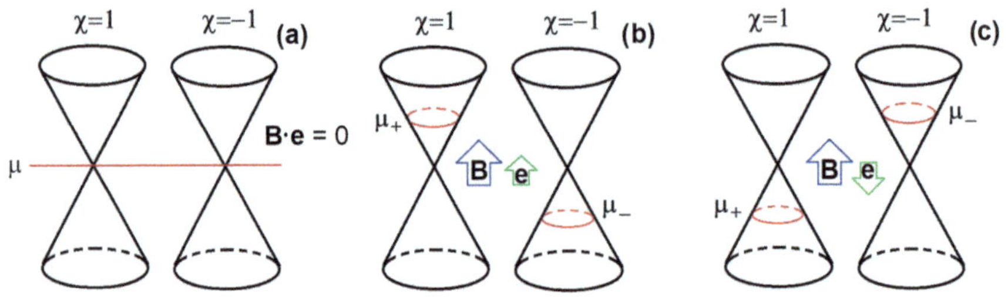

Fig. 1.13 Schematic illustration for the dynamic chiral anomaly: the chemical potential of the Weyl bands oscillate following the time-dependent electric field. The phase shift between the time-dependent μ and $e(t)$ is set by the band chirality

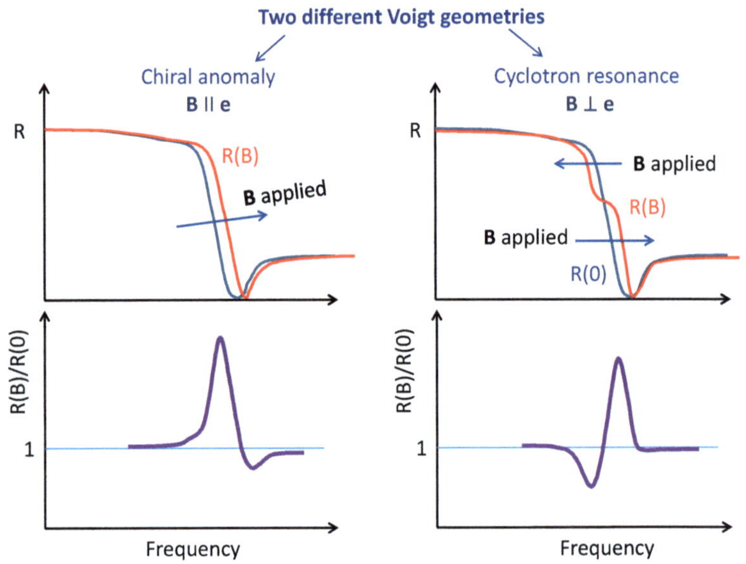

Fig. 1.14 The dynamic chiral anomaly (shift of the plasma frequency) as detected by optics in comparison with the plasma-edge splitting due to a cyclotron resonance. Left panels: the ordinary Voigt geometry, $\mathbf{e} \parallel \mathbf{B}$. Right panels: the extraordinary Voigt geometry, $\mathbf{e} \perp \mathbf{B}$. Top panels: optical reflectivity with a plasma edge for zero (the blue curves) and non-zero (the red curves) external magnetic fields. Bottom panels: the R(B)/R(0) ratios. The horizonal scale on all panels is frequency

for the cones with opposite chiralities. Important is that the average[7] *absolute value* of μ is not zero now. This change of the average chemical potential can be traced in the optical spectra, as we now have more free carriers effectively. One can show that for doped Weyl semimetals the effect will be the same: an effective increase of the carrier density seen as a blue shift of the plasma edge [111, 114].

[7] The averaging should be made on the time scales lager than the wave period, i.e., continuous-wave optics should be utilized for the detection of this effect.

This blue shift of the plasma frequency manifest itself, e.g., in the optical reflectivity, see Fig. 1.14. Here, we sketch the reflectivity of a Weyl semimetal for two different Voigt geometries, $\mathbf{e} \parallel \mathbf{B}$ and $\mathbf{e} \perp \mathbf{B}$. Following [115], we will sometimes refer to them as to "ordinary" and "extraordinary" in the course of the book. For usual materials without chiral electronic bands, no changes in the optical spectra are expected upon application of external magnetic fields in the ordinary Voigt geometry, as there is no Lorentz force for $\mathbf{e} \parallel \mathbf{B}$. The dynamic chiral anomaly is unique in this sense, cf. upper left panel of Fig. 1.14. In the extraordinary Voigt configuration, a cyclotron resonance (CR) can often be seen ([115] and upper right panel of Fig. 1.14).

In practice, the ratio of the reflectivity in an applied magnetic field $R(B)$ to the zero-field reflectivity $R(0)$ is measured in experiment. As one can see from the bottom panels of Fig. 1.14, these ratios look rather different from each other for the two Voigt confirmations in the case of a Weyl semimetal. The chiral anomaly manifests itself as a peak appearing above the unity line in $R(B)/R(0)$, while the CR-induced plasma-edge splitting leads to an oscillating feature in this ratio: $R(B)/R(0)$ goes first down and then up as frequency increases. Observing the bumps in the $R(B)/R(0)$ spectra near the plasma-edge frequencies in the ordinary Voigt geometry is therefore a direct indication of the dynamic chiral anomaly. Pioneering experiments on this subject have been performed by Levy et al. [111].

Chapter 2
Nodal-Line Semimetals

Succeeding the intense investigations of the linearly dispersing energy bands in two-dimensional graphene, nodal-line semimetals (NLSMs) were predicted in 2011 [5]. In these topological semimetals, 2D electronic bands with linear dispersion (Dirac bands) cross each other along continuous lines (loops) in reciprocal space [22, 116–129]. In general, the topology of the nodal lines within a BZ may be very complex; e.g., nodal lines may be linked and knotted in different ways [130–132]. Important is that 2D Dirac electrons exists in the 3D bulk of a NLSM. Such 3D materials with 2D Dirac electrons (i.e. the 3D analogues of graphene) are supposed to demonstrate a number of unusual electronic properties that can potentially be useful for applications [5, 116, 117]. In this book, we concentrate on ZrSiS – a compound, which allows perhaps the most elaborative study of NLSMs, even though it doesn't formally belong to this class of materials: the nodal line in ZrSiS gapped. However, the size of the gap is very small, as discussed below in this Chapter, and, most importantly, the band structure of ZrSiS doesn't possess any bands except the (slightly gapped) Dirac bands, forming the nodal line, for the energies up to 0.5 eV, making this material a model compound for all NLSMs.

2.1 ZrSiS

In recent years, much attention is paid to ZrSiS [22, 125–127] and its structural analogues, such as HfSiS [127, 133], ZrSiTe [128, 134], and CeSbTe [135]. The presence of effectively 2D Dirac bands in ZrSiS and its family is well established by several methods, including ARPES [22, 126–128, 136], Hall measurements [137, 138], and quantum oscillations [134, 136–142], as well as by electronic-structure calculations [22, 126–128, 136]. These studies demonstrate that ZrSiS possesses two types of line nodes. The line nodes of the first type are situated far away ($\sim$0.7 eV)

A. V. Pronin, *Linear Electrodynamic Response of Topological Semimetals*,
Springer Series in Solid-State Sciences 199,
https://doi.org/10.1007/978-3-031-35637-7_2

from the Fermi level; we dub them as high-energy nodes. Spin-orbit coupling opens a gap along certain portions of this nodal line [22, 127]. The second-type line nodes appear close to the Fermi level (low-energy nodes), but are believed to be fully gapped due to SOC, similarly to such NLSM candidates as Cu_3PdN [122] and $SrIrO_3$ [121]. The gap, however, is calculated to be very small, of the order of 10 meV [22]. Such a small value has indeed been confirmed by recent ARPES measurements [127], although the resolution was not sufficient to accurately determine the gap size. At higher energies (up to a few hundreds meV), the linearity of the line-node Dirac bands in ZrSiS remains uncompromised [22, 127]. These low-energy nodal line is in focus of the present study. On ZrSiS, we performed broadband optical-conductivity measurements in zero magnetic field, as well as magneto-optical measurements in the far-infrared region. The results of these studies are published in [143, 144].

2.1.1 *Broadband Spectroscopy*

Here, we discuss the observation of frequency-independent optical conductivity in ZrSiS [143]. This evidences the existence of quasi-2D Dirac states and a quasi-2D electronic ac transport in this material. From our optical measurements, we extract the length of the nodal line (the node is understood here as the Dirac point of the gapped Dirac band) and estimate the size of the gap in this band.

2.1.1.1 Experimental Details

The investigated single crystals were grown by loading equimolar amounts of Zr, Si, and S together with a small amount of iodine in a sealed quartz tube, which was kept at 1100 °C for 1 week. A temperature gradient of 100 °C was applied and the crystals were collected at the cold end of the tube. The crystal structure (tetragonal, space group *P4/mmm*) was confirmed with X-ray and electron diffraction similarly to [22].

The optical reflectivity $R(\nu)$ was measured at 10–300 K over a broad frequency range from ~50 to 25000 cm^{-1} using commercial Fourier-transform infrared spectrometers. All measurements were performed on freshly cleaved (001) surfaces. In accordance with the tetragonal structure, no in-plane optical anisotropy was detected. At low frequencies, an in-situ gold evaporation technique was utilized for reference measurements, while above 1000 cm^{-1} gold and protected silver mirrors served as references. The high-frequency range was extended by room-temperature ellipsometry measurements up to 45 000 cm^{-1} in order to obtain more accurate results for the Kramers-Kronig analysis. The Kramers-Kronig analysis was made involving the x-ray atomic scattering functions for high-frequency extrapolations [145] and dc-conductivity values, $\sigma_{dc}(T)$, and the reflectivity-fitting procedure [146] for zero-frequency extrapolations. Important to note, that our optical measurements reflect the bulk material properties, since the skin depth is above 40 nm for any measurement frequency.

2.1.1.2 Results and Discussion

The measured frequency-dependent reflectivity $R(\nu)$ is shown in Fig. 2.1 for selected temperatures. Above 1000 cm^{-1}, the temperature has only a minor influence on the spectra. In the low-frequency range, the reflectivity is rather high (above 99%), in agreement with the very low dc resistivity [137, 139]. The results of the Kramers-Kronig analysis are shown in Figs. 2.2, 2.3 and 2.4 in terms of the real and imaginary parts of optical conductivity, as well as the real part of permittivity.

An important result of this work is presented in Fig. 2.2: the real part of optical conductivity is almost frequency-independent, $\sigma_1(\omega) = \sigma_{\text{flat}} \approx 6600\ \Omega^{-1}\text{cm}^{-1}$, in the range from 250 to 2500 cm^{-1} [30 – 300 meV] basically at all temperatures investigated (at $T \geq 100$ K, the flat region starts at a bit higher frequencies because of a rather broad free-electron contribution). Such frequency-independent behavior of $\sigma_1(\omega)$ is similar to what has been predicted [99] and observed [100] in graphene and matches the theory for the optical response due to transitions between the 2D Dirac states in NLSMs, see this chapter. As discussed there, σ_{flat} should be related to the length of the nodal line k_0. For a circular nodal line and a particle-hole symmetric Dirac band, (1.37) is valid, if the electric-field component of the probing radiation is perpendicular to the plane of the nodal circle [84, 85, 102]. In ZrSiS, the low-energy nodal line is not circular and not even flat; instead the BZ contains a 3D "cage" of nodal lines, see Fig. 2.1 and [22]. Thus, a straightforward application of (1.37) is not rigourously validated. Nevertheless, having no better model at hand, we use this equation for a rough estimate of $k_0 = 4.3$ Å^{-1}. This value seems to be reasonable: according to the band-structure calculations, the total lengths of the nodal

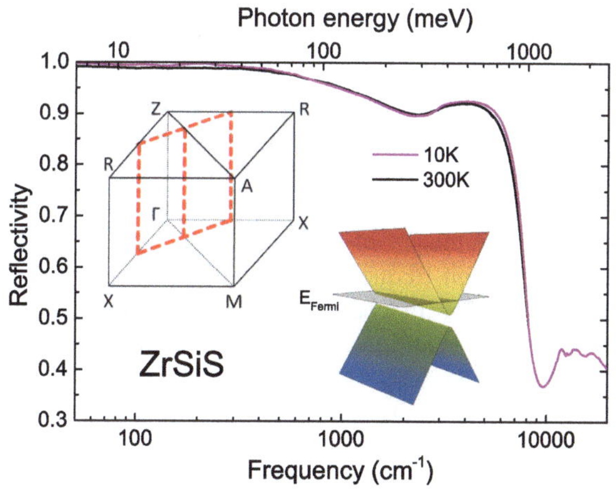

Fig. 2.1 Frequency-dependent reflectivity of ZrSiS at 10 and 300 K. For intermediate temperatures the reflectivity curves lie in between, and are not shown for clarity. No temperature dependence is seen above ~8000 cm^{-1}. The sketches show the position of the low-energy nodal line (red dashed line) in the BZ and the Dirac bands near the Fermi level. Adapted with permission from [143]. Copyright (2017) by the American Physical Society

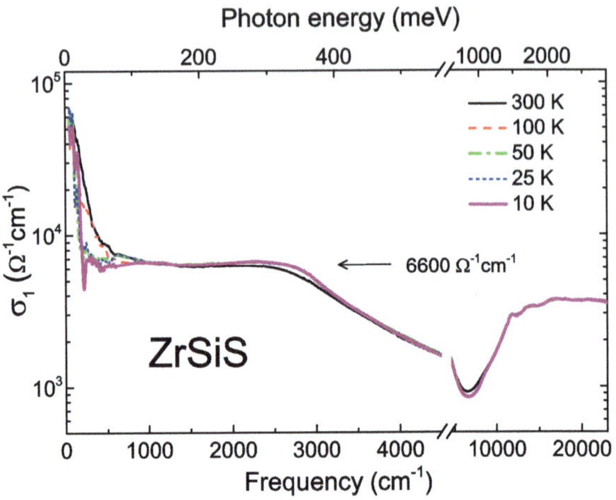

Fig. 2.2 Real part of the optical conductivity of ZrSiS as a function of frequency. Reproduced with permission from [143]. Copyright (2017) by the American Physical Society

line projections on the [100] and [001] directions are about 3.5 and 6 Å^{-1} per BZ, respectively.

At $\nu > 3000$ cm^{-1}, $\sigma_1(\omega)$ is not frequency-independent anymore—it decreases with frequency. At even higher frequencies, $\sigma_1(\omega)$ starts to rise as ω increases. The overall $\sigma_1(\omega)$ spectrum demonstrates here a so-called U-shape behavior. After our paper on ZrSiS was published, similar behavior was also observed in a number of related compounds—ZrSiSe, ZrGeS, and ZrGeTe [147] and reproduced in band-structure-based calculations for all four materials [147, 148].

The interpretation of the high-energy upturn of the U-shaped conductivity is rather straightforward – it is due to transitions between almost parallel bands near the X and R points of the BZ of these compounds. According to the study by Ebad-Allah et al. [147], the low-energy part of the U-shaped conductivity, where σ_1 is roughly proportional to $1/\omega$, can be interpreted as being due to an "effective nodal plane" (see (1.33) for $z = 1$ and $d = 1$). In other words, there is a further electronic-band dimensionality reduction due to the quasi-2D (band) structure of these compounds: the dispersion is linear only in one k-space direction, whereas it is almost absent along the nodal line, as well as in the out-of-plane direction. This nodal-plane picture can only work for relatively high frequencies, as otherwise the band structure cannot be approximated as quasi-2D. Indeed, a basically frequency-independent $\sigma_1(\omega)$ is observed at low frequencies in all four materials, albeit the frequency span of the flat conductivity is largest in ZrSiS.

These flat areas in the interband conductivity, $\sigma_1^{\mathrm{IB}}(\omega)$, cannot be accurately reproduced by the available calculations [147, 148]. Instead, the calculated low-energy $\sigma_1^{\mathrm{IB}}(\omega)$ is found to increase with frequency. Habe and Koshino [148] suggested that the observed flat conductivity might be a cumulative effect of the increasing interband conductivity and a decreasing Drude contribution. However, the experimental data do not support this explanation. As shown in Fig. 2.2, the Drude term is very narrow and does not overlap with the flat region. Also, it is the interband conductivity

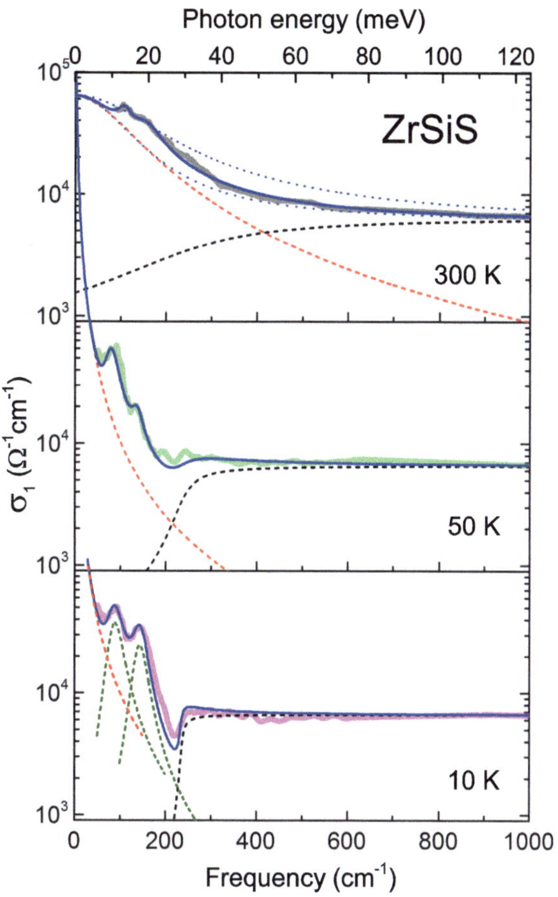

Fig. 2.3 Examples of the optical conductivity fits for ZrSiS at low frequencies. Thick solid lines are experimental data; thin solid lines are total fits; dashed lines represent fit contributions of the Drude (red), Pauli-edge (black), and Lorentzian (olive, shown only for T = 10 K) terms. Thin dotted (blue) lines represents attempts to fit the 300-K data without the Lorentzians. Reproduced with permission from [143]. Copyright (2017) by the American Physical Society

(i.e., the conductivity after subtraction of the Drude modes), which shows the almost flat regions at low energies in the study by Ebad-Allah et al.

It would be of interest to detect the transitions between the high-energy Dirac bands above approximately 1.3 eV = 10 500 cm^{-1}, but we do not see any clear signatures of such transitions: as discussed above, these Dirac bands are partly gapped by SOC and for bands of such complex shapes the optical conductivity is not expected to be flat or linear in frequency. Additionally, non-Dirac bands above the Fermi level also contribute to the absorption processes at these frequencies, as one can see from the band structure calculations of [22, 127].

At $\nu < 500$ cm^{-1}, $\sigma_1(\omega)$ also deviates from the flat behavior, as can be seen in Figs. 2.2 and 2.3, and exhibits several features. The contribution of free carriers (the Drude response) is present at all temperatures, but best seen at 100 and 300 K. A free-electron component is expected in the optical response because the Fermi level in ZrSiS is slightly above the nodal line [22, 127]. As the temperature drops, the Drude band narrows, revealing two distinct modes at around 100 and 150 cm^{-1}. A

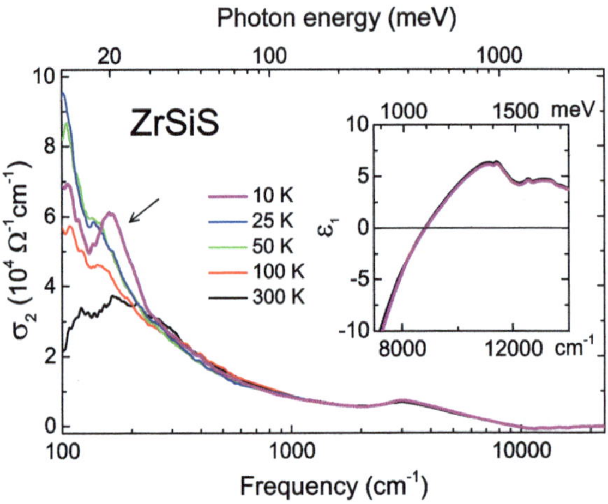

Fig. 2.4 Main frame: imaginary part of the optical conductivity in ZrSiS. Arrow indicates the position of the Pauli edge at 10 K. Inset: real part of the dielectric constant near the plasma frequency. Reproduced with permission from [143]. Copyright (2017) by the American Physical Society

brief discussion on the possible origins of these modes is given below. The narrowing of the Drude band reflects the strong suppression of the dc resistivity [137, 139] and of the carrier scattering rate (or, more accurately, the momentum-relaxation rate) as $T \to 0$.

At the lowest temperatures, $\sigma_1(\omega)$ develops a minimum around 200 cm^{-1} (bottom panel of Fig. 2.3). We relate this dip and the corresponding feature in $\sigma_2(\omega)$, see Fig. 2.4, to the Pauli blocking of the transitions in the 2D band. Such features in $\sigma_2(\omega)$, related to the position of the Fermi level, are well known in semiconductors [79] and have recently been discussed in relation to graphene and Dirac/Weyl semimetals [149–151]. As already mentioned, band-structure calculations and ARPES measurements locate the Fermi level in ZrSiS in the upper (conduction) Dirac band [22, 127], see the sketch in Fig. 2.1. Thus, the Pauli edge must be seen in the interband portion of optical conductivity. An onset of the interband transitions shows up at the frequency equal to $\max\{\Delta, 2\mu\}$ [85, 102, 149–151], with Δ being the band gap and μ the position of the Fermi level relative to the Dirac point. Thus, (1.37) can be modified to:

$$\sigma_1(\omega) = \frac{e^2 k_0}{16\hbar} \times \theta(\hbar\omega - \max\{\Delta, 2\mu\}), \tag{2.1}$$

where $\theta(x)$ is the Heaviside step function. From (2.1) and the bottom panel of Fig. 2.3, one can conclude that $\max\{\Delta, 2\mu\}$ must be smaller than approximately 250 cm^{-1} or some 30 meV and, hence $\Delta < 30$ meV.

This estimate of the upper limit of Δ from optical data is in good agreement with the value obtained from band-structure calculations (15 meV, [22]). The ARPES value of 60 meV [127] likely overestimates the gap due to the limited resolution. The fact that the relative position of the Dirac points and the Fermi level is slightly $\mathbf{k}$-dependent [22, 127] might lead to a broadening of the optical Pauli-edge feature observed. Importantly, even if the Fermi level appears within the gap for some values of **k**, our conclusion on the upper limit of Δ still holds.

Let us now turn to the modes at 100 and 150 cm^{-1} and consider possible scenarios for their origin. (i) Phonons. It is extremely unlikely that the observed modes originate from phonons as such, because the frequencies of the modes do not agree with calculations of the infrared-active phonons [152] and because the modes are unrealistically strong for any phonon except soft modes in ferroelectrics. However, no ferroelectricity is reported/expected in ZrSiS and our modes do not shift with temperature, as it would be the case for ferroelectric soft modes. (ii) Bulk electron localization. A disorder-induced localization of bulk electrons seems also rather unlikely, since the material is very clean [142]. Its dc resistivity monotonously decreases with temperature and is extremely low at $T \to 0$ [137, 139]. (iii) Surface contributions. Although our skin depth is above 40 nm for any measurement frequency, some surface contributions cannot be fully excluded. Very recently, extremely strong surface states in ZrSiS have been theoretically predicted and observed by ARPES measurements [153]. However, no surface contributions are seen in dc-conductivity measurements on microstructure samples with different thicknesses. Appearance of inhomogeneities in the electron surface density, possibly with electron localization, can reconcile this fact and the results of [153], as well as provide an explanation for our low-frequency modes. (iv) Excitons. At first glance, it does not seem plausible that excitons are the reason for the observed features, because of the strong screening effects from free carriers. Nevertheless, if we calculate the exciton binding energy E_b within the simplest hydrogenic exciton model [154] using the effective mass of the carriers in the low-energy 2D Dirac band, $m^* = 0.025m_e$ [142], and an estimate of the static lattice dielectric constant, $\varepsilon_0 \sim 7.5$,[1] we obtain $E_b \sim 100$ cm^{-1}. Excitons with such E_b are supposed to be seen in the spectra at $\Delta - E_b \leq 150$ cm^{-1} ($\Delta \leq 250$ cm^{-1}), i.e. right in the frequency range of our modes. One can propose that the excitons may appear on the surface, where free-electron screening is weaker. Furthermore, large strengths of the observed modes in this case can be quantitatively explained by bounding between excitons and surface inhomogeneities [155]. Further studies are necessary to fully clarify the nature of the low-energy absorption peaks in ZrSiS.

To get some more quantitative estimates of the parameters determining the optical response, we fit the optical conductivity with a model consisting of a Drude term, two Lorentzians, and a term describing the Pauli edge. Scattering and other processes, leading to broadening of the sharp step in (2.1), may be taken into account by replacing the Heaviside function with

$$\frac{1}{2} + \frac{1}{\pi} \arctan \frac{\omega - \max\{\Delta, 2\mu\}/\hbar}{\Gamma}, \tag{2.2}$$

[1] This estimate can be obtained by subtracting the Drude contribution from the measured spectra of $\varepsilon_1(\omega)$.

for example, where Γ represents a broadening parameter due to **k**-dependent gap, impurity scattering, or temperature (cf. (1.36)). For $T = 10$ K, reasonable fits can be obtained with a very sharp Pauli edge, i.e. with Γ of a few cm^{-1}. We set $\Gamma = k_B T/\hbar$ for all temperatures, since smaller values seem not to be physical. This yields $\Gamma = 7$, 35, and 210 cm^{-1} for 10, 50, and 300 K, respectively, see Fig. 2.3.

In all our fits we keep the zero-frequency limit of the Drude term equal to σ_{dc} at all temperatures. Owed to the broad Drude tail, the description of the 300-K data is straightforward. It provides the momentum-relaxation rate of free carriers, $\gamma = 1/(2\pi\tau) = (120 \pm 10)$ cm^{-1} (τ is the corresponding relaxation time), and a plasma frequency, $\omega_{pl}/2\pi = (24000 \pm 1000)$ cm^{-1}.[2] On the other hand, the screened plasma frequency, $\omega_{pl}^{scr} = \omega_{pl}/\sqrt{\varepsilon_\infty}$ (ε_∞ is the contribution of the higher-frequency optical transitions to ε_1), can be directly determined from optical measurements as the zero-crossing point of $\varepsilon_1(\nu)$ [75]. We find ω_{pl}^{scr} to be temperature independent and situated at 8900 cm^{-1}, cf. the inset of Fig. 2.4. Hence, $\varepsilon_\infty = (\omega_{pl}/\omega_{pl}^{scr})^2 \approx 7$, which is in good agrement with the optical measurements, presented in same figure.

As one can see from Figs. 2.2, 2.3 and 2.4, the Drude term becomes narrower as $T \to 0$. At low temperatures, γ is below our measurement window and our fits thus might become ambiguous. To avoid this, we first tried to keep the plasma frequency of the Drude term constant as a function of T; but this turned out to be unsatisfactory. Some spectral weight had to be redistributed between the Drude term and the Lorentzians. Nevertheless, we tried to have this spectral weight transfer as small as possible and the total plasma frequency of the three terms (Drude plus two Lorentzians) to be temperature-independent in accordance with the temperature-independent ω_{pl}. Examples of the fits obtained in this way are shown in Fig. 2.3. At $T \leq 50$ K, we found that $\gamma \approx 2$ to 2.5 cm^{-1} and the momentum-relaxing τ is 2.1 to 2.7 ps.

Interestingly, at low temperatures the momentum-relaxation length, $\ell_{\mathrm{mr}} = v_F\tau$, obtained from our estimate of τ, becomes macroscopically large. Using $v_F = 5 \times 10^5$ m/s as an average Fermi velocity in the low-energy Dirac bands [137], we obtain $\ell_{\mathrm{mr}} \geq 1$ μm for $T \leq 50$ K. This implies that the hydrodynamic behavior of electrons, reported recently in clean samples of graphene [156, 157] and the Weyl semimetal WP_2 [158], might also be realized in ZrSiS. This proposition seems reasonable, because only linear bands with highly mobile carriers (typical mobilities are 10^3 to 10^4 cm^2/Vs [137, 138, 142]) cross the Fermi level in ZrSiS.

[2] Note, even at $T = 300$ K, the Lorentzians in our model improve the fit, see Fig. 2.3a.

2.1.2 *Magneto-Optical Response*

Magneto-optical spectroscopy[3] is a powerful tool to investigate electronic properties of nodal semimetals and narrow-band-gap materials [103, 159–162]. For example, this method enables experimental verification of the electronic band structure by tracing the optical transitions between the magnetic-field-induced Landau levels (LLs). This approach is particularly relevant at low energies, where other experimental techniques often lack accuracy and resolution. Here, we present the far-infrared magneto-optical investigations of ZrSiS, reported by us in [144]. We found that Dirac quasiparticles fully dominate the ac (magneto)transport in ZrSiS: both inter- and intra-band optical transitions demonstrate a square-root dependence on magnetic field, typical for such bands. All measured magneto-optical spectra can be well described by a simple model of gapped Dirac bands with a single (i.e. **k**-independent) gap of 26 meV (210 cm^{-1}).

2.1.2.1 Experimental Details

The sample studied by magneto-optics was the same single crystal as the one used in the zero-field optical measurements described above. The optical reflectivity spectra were collected from (001) surfaces (the in-plane response) utilizing a home-build magneto-optical setup connected to a commercial Fourier-transform infrared spectrometer [163]. The spectra were recorded for photon energies between ~5 and 75 meV (40–600 cm^{-1}) in magnetic fields up to 7 T at 10 K. The Voigt geometry was chosen as the measurement configuration [115]. The spectra have been obtained with two linear polarizations, $\mathbf{e}||\mathbf{B}$ and $\mathbf{e} \perp \mathbf{B}$ (here, **e** is the electric component of the probing radiation). In the $\mathbf{e}||\mathbf{B}$ polarization, no field-induced changes were detected. Hereafter in this Chapter, we describe our results obtained for $\mathbf{e} \perp \mathbf{B}$. A simultaneous change of the **B**-field direction and the light polarization (keeping the angle between them fixed) in respect to the crystallographic direction within the (001) plane did not affect the spectra due to the tetragonal crystal symmetry within this plane. In order to obtain the optical conductivity, the measured reflectivity spectra were merged with the zero-field reflectivity discussed above at higher energies and the Kramers-Kronig analysis was performed using x-ray atomic scattering functions [145] and Drude-Lorentz fits as extrapolations at high and low frequencies [146], respectively.

---

[3] In magneto-optics, the frequency is traditionally expressed in meV, while in the broadband optics cm^{-1} are usually used. We stick to these customs in this book and present our results accordingly. For the most important findings, we use both units. For rough estimates, one can always keep in mind that 1 mev is roughly 10 cm^{-1} (more accurately, 1 meV corresponds to 8.065541 cm^{-1}.)

2.1.2.2 Results and Discussion

The measured reflectivity is given in Fig. 2.5. To demonstrate the B-field-induced changes more clearly, we plot the relative reflectivity, $R(B)/R(B = 0)$. The corresponding spectra of the real part of optical conductivity, normalized by its zero-field value, $\sigma(B)/\sigma(B = 0)$, are given in Fig. 2.6a. The spectra between ~17 and 24 meV are largely affected by noise in our setup and therefore eliminated from the discussion. Above 50 meV no field-induced features appear; we chose the energy scale on the plots accordingly.

The optical spectra demonstrate three field-induced features: (i) a suppression of reflectivity and optical conductivity at the lowest frequencies (marked with the red arrows in both figures); (ii) a feature that grows in intensity with increasing B, but remains its frequency position constant at around 15 meV [the gray triangle in

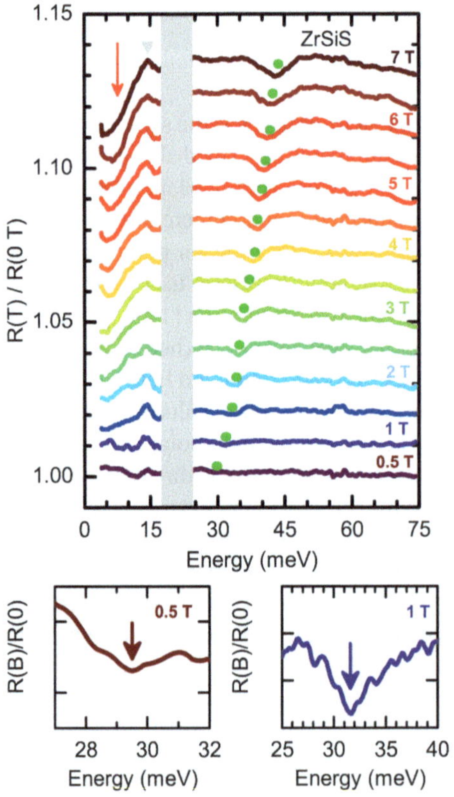

Fig. 2.5 Magneto-reflectance spectra of ZrSiS normalized to zero-field reflectivity [144]. The spectra at various fields are shifted by 0.01 with respect to the previous one for clarity. The shaded area is the range not accessible for our measurement setup. Enlarged spectra near one of the absorption modes for 0.5 T and 1 T are given in the bottom panels. Reprinted from [144] under CC-BY 4.0 licence. Copyright (2019) by the authors

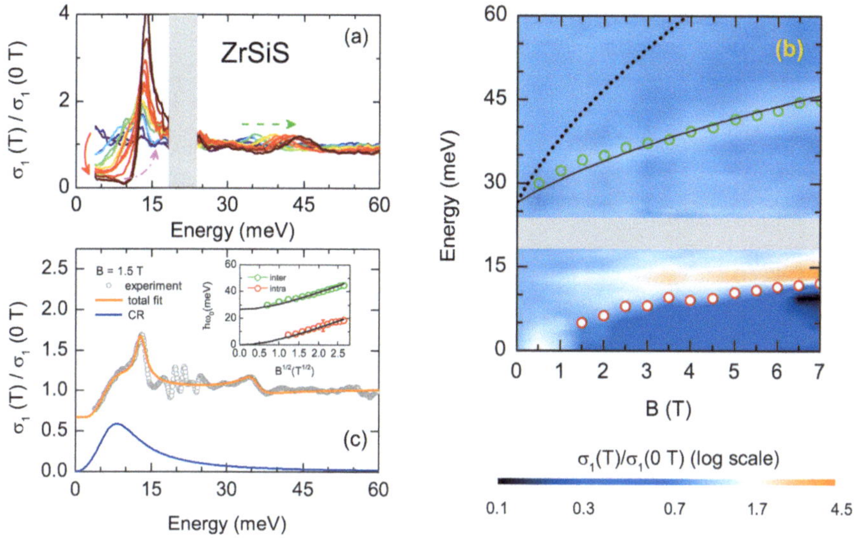

Fig. 2.6 Panel **a**: Relative optical conductivity of ZrSiS versus frequency for all studied fields. Panel **b**: Same data as a false-color plot. Lines and dots correspond to different inter-LL transitions as discussed in the text. Panel **c**: An example of the Lorentz-Drude fit to the optical conductivity data at 1.5 T. The structures between approximately 17 and 24 meV are noise (corresponding to the shaded areas in the other panels). The inset shows the energy positions ($\hbar\omega_0$) of the inter- and intra-band transitions determined from the conductivity data and fitted with (2.4). Note the $\sqrt{B}$ horizontal scale. Reprinted from [144] under CC-BY 4.0 licence. Copyright (2019) by the authors

Fig. 2.5 and the strongest peak in Fig. 2.6a]; and finally, (iii) a feature around 30–45 meV seen at all the fields and marked with the green circles in Fig. 2.5 and the green arrow in Fig. 2.6a demonstrating the frequency shift of the feature as B increases. In Fig. 2.6b, a false-color plot of the relative optical conductivity as a function of energy and magnetic field is shown. The three field-induced features are seen in this plot as changes of color. The features (i) and (iii) are further emphasized by the open symbols obtained as discussed below.

Before going to the discussion, let us recall that for gapped Dirac bands the LL spectrum reads as [162, 164]:

$$E_{\pm n} = \pm\sqrt{2e\hbar \mid n \mid B v_F^2 + \left(\frac{\Delta}{2}\right)^2}, \quad (2.3)$$

where n is the LL index, v_F is the (average) Fermi velocity, and Δ is the gap between conduction and valence bands. The LLs in the conduction and valence bands correspond to the "+" and "−" signs in (2.3), respectively. The spectrum of (2.3) is shown in the upper panel of Fig. 2.7. For the allowed inter-LL transitions, $|n|$ can be changed by ± 1. Hence, the transition energy can be written, e.g., as:

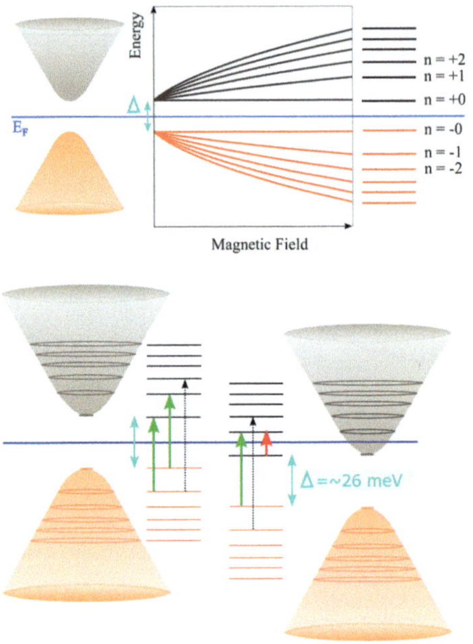

Fig. 2.7 Upper panel: Schematic representation of Landau levels appearing in a gapped Dirac band. Bottom panels sketch the 2D Dirac band in ZrSiS. Two cuts perpendicular to the nodal line are shown: along some portions of the gapped nodal line, the Fermi level is in the gap (left picture), while in the other portions (right picture) it is in the conduction band. The cyan solid arrow represents the determined single band gap. Other solid arrows depict the observed transitions: the intra-band (CR) transition $+0 \rightarrow +1$ is shown as a red arrow, while the interband transitions ($-0 \rightarrow +1$ and $-1 \rightarrow +0$) are given as green arrows. The dotted arrows show other possible interband transitions, which are not observed in this work. The color code for the arrows is consistent with the one used to mark the transition modes in Figs. 2.5 and 2.6. Reprinted from [144] under CC-BY 4.0 licence. Copyright (2019) by the authors

$$E_t = \sqrt{2e\hbar(\mid n \mid +1)Bv_F^2 + \left(\frac{\Delta}{2}\right)^2} \pm \sqrt{2e\hbar \mid n \mid Bv_F^2 + \left(\frac{\Delta}{2}\right)^2} \qquad (2.4)$$

with the plus sign corresponding to interband and the minus sign to intraband transitions. These transitions are schematically depicted in the bottom part of Fig. 2.7. Two situations are considered: with E_F in the gap (left panel) and in the conduction band (right panel). Both situations are relevant for ZrSiS: it is well known that the Fermi level crosses electronic bands along some portions of the nodal line, while it is within the gap elsewhere [22]. In the following, we argue that this simple model can account for all field-dependent features observed by us in the magneto-optical spectra of ZrSiS.

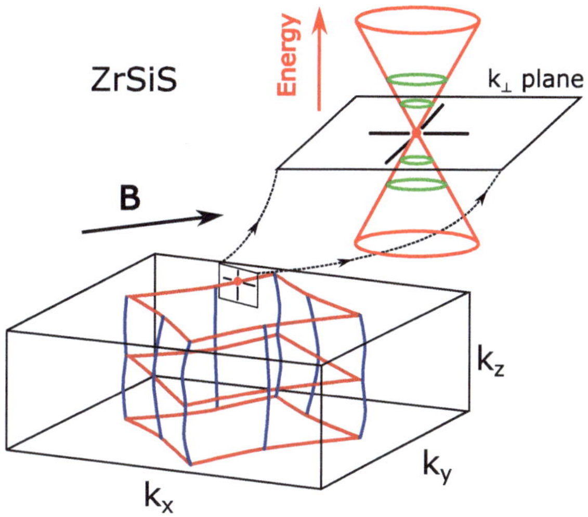

Fig. 2.8 Schematics of the nodal line in ZrSiS. The 2D Dirac band with its Landau levels formed for the in-plane orientation of **B** is sketched in the upper right corner ($k_\perp$ stands for the k-directions perpendicular to the nodal line). The band gap is not shown for simplicity. Reprinted from [144] under CC-BY 4.0 licence. Copyright (2019) by the authors

It is worth noting here that optical-conductivity measurements are not k-sensitive and probe the entire BZ. In the present study, we mostly probe the in-plane parts of the nodal line, because the external magnetic field is applied parallel to the (001) plane and hence the closed cyclotron orbits are formed for the carriers from the bands forming the in-plane portions of the nodal line, see Fig. 2.8. (Small contributions from the out-of-plane parts can also be present due to their corrugations.)

We start from the lowest energy feature (i). To make it more visible, we added the red open circles to Fig. 2.6b. Their positions are determined as the unity-crossing points of $\sigma(B)/\sigma(B = 0)$, cf. Fig. 2.6a. The feature (i) starts to appear at 1.5 T and seems to extrapolate to zero as $B \to 0$. By looking on the field evolution of this feature in Fig. 2.6a, it can clearly be related to depletion of the Drude response: the loss of Drude spectral weight at low energies is expected to be transferred to a cyclotron-resonance (CR) mode [160], as indicated by the dash-dotted pink arrow in Fig. 2.6a. In our measurements the cyclotron resonance is superimposed on the field-independent feature (ii). In order to extract the CR parameters, we performed a Drude-Lorentz fit [75] that takes into account the CR itself, the leftover Drude component, the field-independent resonance, and the high-energy mode corresponding to feature (iii), the latter is taken into account only for completeness. In Fig. 2.6c, we present an example of such a fit for 1.5 T. This way we can extract the field dependence of the cyclotron resonance. In the inset of the Fig. 2.6c we plot the obtained energies of the cyclotron resonance as a function of B field. The linear behavior on the $\sqrt{B}$-scale proves the relativistic nature (linear bands) of the carriers responsible for this resonance and can be nicely fitted by (2.4) using $n = 0$ and the minus sign corresponding to the intraband transition (the red arrow in Fig. 2.7).

Now, we turn to the feature (iii). At low B, it extrapolates to the gap detected in our zero-field measurements and it can be assigned to the transitions between LLs in the valence and conduction bands. In Fig. 2.6b, we plot the peak positions of $\sigma(B)/\sigma(B=0)$ as open green symbols. Remarkably, we can fit the symbols with (2.4) using the same Δ and v_F as we used to fit the CR mode. Only the sign between the addends should be changed (minus to plus). Thus, the observed mode (iii) corresponds to the interband transitions involving the $n=0$ LLs, i.e., the transitions from $n=-1$ to $n=+0$ or from $n=-0$ to $n=+1$ (green arrows in Fig. 2.7).

The fits to the B-dependent modes (i) and (iii) shown in Fig. 2.6b, c as black solid lines are all performed using the single values of $\Delta = 26$ meV and $v_F = 3\times10^5$ m/s. This strongly indicates that the distribution of these parameters along (the in-plane portions of) the nodal line in the BZ is rather narrow, as optical probes may only provide momentum-averaged quantities. Had Δ and v_F possessed strong **k**-dispersions, the field-induced magneto-optical features would be extremely broad, if detectable. As already noticed, the gap identified is well within the margins provided by calculations [22], ARPES [127], and our optical data in zero magnetic field. The same is true for the obtained Fermi velocity, cf. [139, 165].

In Fig. 2.6b, we also show the position expected for the inter-LL transitions from $n=-2$ to $n=+1$ and from $n=-1$ to $n=+2$ (dotted line, see also dotted arrows in Fig. 2.7). These, as well as other transitions between the LLs with larger n, are not detectable in our spectra. This situation is quite common: the line broadening increases with energy [159, 166] making the line detection difficult.

Finally, let us turn to the field-independent feature (ii), which is seen around 15 meV. The position of this mode does not show any appreciable frequency shift, but the mode strength increases with magnetic field. A simple explanation of this mode would be an in-gap impurity resonance. However, the studied ZrSiS sample is very clean: the mean free path extracted from our optical measurements on this sample is of the order of 1 μm at low temperatures, see above. Thus, this explanation seems not to be very likely. Alternatively, surface states [153] might be relevant for the formation of this mode. A more appealing option is of bulk and intrinsic origin. We note that a low-energy mode with a very similar behavior in magnetic field (constant energy position and strength increasing with B) was observed, but remained unexplained, in another gapped NLSM, $NbAs_2$ [103]. It would be tempting to assign both modes to the B-independent resonance, recently theoretically proposed to be a hallmark of the NLSM state [167]. However, the model used in this reference (a single nodal loop formed by crossing cones) is far too simple and not directly applicable to ZrSiS or $NbAs_2$: both compounds possess complex and extended nodal lines. Thus, we call for more theoretical studies in this direction.

2.1.3 Conclusions

Summarizing the results and analysis of our broadband conductivity measurements, we conclude that the real part of the optical conductivity of ZrSiS is independent on frequency in the range from 250 to 2500 cm^{-1} (30–300 meV). Our observations are supported by the theoretical predictions for the optical response of NLSMs, and constitute and independent confirmation for 2D Dirac bands in ZrSiS near the Fermi level. The characteristic features of the Pauli edge, appearing in the low-frequency spectra, provide the upper limit (250 cm^{-1}, 30 meV) for the gap between the 2D Dirac bands. The momentum-relaxation length is at the micrometer scale at $T \leq 50$ K. Overall, our optical measurements reveal that ZrSiS is a gapped line-node semimetal with the electronic properties determined primarily by 2D Dirac electrons with rather slow momentum relaxation at low temperatures.

In turn, our far-infrared magneto-optical study of ZrSiS reveals three features developing in applied magnetic field. The energy position of two of them scales as a square root of B. A simple model of a single gapped Dirac band provides adequate explanation for the field evolution of these features, the low-energy feature being associated with intra-band (cyclotron resonance) absorption, while the high-energy one appearing due to interband LL transitions. The free parameters of this model (the band gap and the Fermi velocity) are defined as $\Delta = 210 \pm 15$ cm^{-1} $= (26 \pm 2)$ meV and $v_F = (3.0 \pm 0.2) \times 10^5$ m/s, which can be considered as averaged values over the entire in-plane part of the nodal line. Both values are well consistent with the ones obtained by other methods and reported in literature. The fact that such a simple model accounts for the complete magneto-optical spectrum indicates that Δ and v_F do not vary appreciably along (the in-plane portion of) the nodal line. This also evidences that other bands do not contribute to the corresponding ac transport. Possible origins of the third observed mode are discussed and more theory output is called for.

Chapter 3
Dirac and Weyl Semimetals

The interest in measurements of the optical conductivity in Dirac and Wey semimetals is triggered by the fact that the interband conductivity in these systems is expected to be proportional to frequency (in the case of perfectly linear Dirac/Weyl cones), see (1.34). Such linearity in $\sigma_1(\omega)$ over a broad frequency range in a 3D system is often considered as a "smoking gun" for Dirac physics regardless the origin of the linear bands. A linear-in-frequency conductivity was observed in the inverted-band semiconductor $Hg_{1-x}Cd_xTe$ with $x = 0.17$ [168]. Timusk et al. [169] claimed the presence of 3D Dirac fermions in a number of quasicrystals (e.g. in $Al_{63.5}Cu_{24.5}Fe_{12}$ and $Al_{75.5}Mn_{20.5}Si_{10.1}$) based entirely on the observation of a linear frequency-dependent conductivity in these materials. Linear-in-frequency $\sigma_1(\omega)$ was also reported in $ZrTe_5$ [170], where the presence of 3D linear bands is supported by transport and angle-resolved photoemission experiments [171].

In this chapter, we present results of our optical measurements for four 3D semimetals. We do observe the linear-in-frequency $\sigma_1(\omega)$, but, as we show in the following chapters, such linearity does not necessarily come from the 3D conical bands. Below, we present our results for two Dirac and two Weyl semimetals. These investigations are published in [172–176].

3.1 The Dirac Semimetal Cd_3As_2

In 2013 Wang et al. [8] predicted 3D topological Dirac points in Cd_3As_2; by now they are well confirmed by ARPES, scanning tunneling spectroscopy, and magneto-transport measurements [12–14, 177–179]. Due to the presence of inversion symmetry in Cd_3As_2, the bands are not spin polarized [180]. The shape of the Dirac bands in Cd_3As_2 is somewhat complicated by the presence of a Lifshitz-transition point [177]; see Fig. 3.2 for a sketch of the Dirac bands.

A. V. Pronin, *Linear Electrodynamic Response of Topological Semimetals*,
Springer Series in Solid-State Sciences 199,
https://doi.org/10.1007/978-3-031-35637-7_3

The goal of our measurements [172] was to provide insight into the Dirac-band dispersion in Cd_3As_2 by means of optical spectroscopy. Previous optical investigations of Cd_3As_2 performed in the 70s and 80s can be divided into two groups. The first one [181–186] deals with the low-energy part of the optical spectra (usually, below some 250 meV, or 2000 cm^{-1}) discussing mostly phonon modes and free-electron Drude-like absorption. Turner et al. [181] identify a very narrow (130 meV) optical gap in the optical absorption. Another group of articles [187–189] mainly discusses absorption features at energies of a few electron volts and their relations to transitions between different (high-energy) parabolic bands. No optical conductivity was derived from these measurements.

The recent recognition of the non-trivial election-band topology of Cd_3As_2 calls for a fresh look onto its optical properties. Here, we present broadband optical investigations of Cd_3As_2. From the reflectivity data, measured from 6.3 meV to 2.7 eV (50–22 000 cm^{-1}), we derive the optical conductivity that is determined by contributions from three different channels: free (itinerant) carriers, phonon modes, and interband transitions. The later start around 200 meV and span up to the highest frequencies available for our measurements. We argue that the low-frequency part of the interband conductivity (below approximately 1 eV) reflects transitions between the Dirac bands in Cd_3As_2. From the $\sigma_1(\omega)$ spectra, the position of the Fermi level is determined, as well as the Fermi velocity of the Dirac particles.

3.1.1 Experiment

Single crystals of Cd_3As_2 have been grown by vapor transport from material previously synthesized in argon flow (with the flow rate of a few cm^3/min) [190]. The temperatures in the evaporation and growth zones are 520 °C and 480 °C, respectively. By subsequent annealing at room temperature the electron mobility is enhanced while at the same time the electron concentration decreases [191]. Well-defined crystals with facets measured a few millimeters in each direction are harvested after 24 hours of the growth process.

The lattice parameters have been obtained by x-ray diffraction: $a = 1.267$ nm and $c = 2.548$ nm, in good agreement with recent measurements reporting a tetragonal unit cell with only light deviations from a cubic structure [180]. Resistivity and Hall measurements provide an electron density of $n_e = 6 \times 10^{17}$ cm^{-3} (roughly independent of temperature), a metallic resistivity, and a mobility of $\mu = 8 \times 10^4$ cm^2/Vs at 12 K.

The investigated Cd_3As_2 single crystal had lateral dimensions of 2.5 mm by 3 mm and a thickness of 300 μm. It was cut out from a larger single crystal. The crystallographic axes of the sample were found by x-ray diffraction. The [001] axis was perpendicular to the sample's largest surface. This surface was polished prior the optical measurements, which were performed for a few linear polarizations. The dc resistivity of this sample was characterized in-plane by standard four-probe method. Both dc and optical measurements revealed an isotropic response with the (001)-plane.

The optical reflectivity was measured from room temperatures down to 10 K with light polarized along different crystallographic directions. The spectra in the far-infrared (50–1000 cm^{-1}) were recorded by a Bruker IFS 113v Fourier-transform infrared spectrometer using an in-situ gold overfilling technique for reference measurements [192]. At higher frequencies (700–22 000 cm^{-1}) a Bruker Hyperion microscope attached to a Bruker Vertex 80v spectrometer was used. Here, either freshly evaporated gold mirrors or coated silver mirrors were utilized as references.

3.1.2 Kramers-Kronig Analysis and its Robustness

For Kramers-Kronig analysis, extrapolations to zero frequency have been made using the Hagen-Rubens relation in accordance with the temperature-dependent dc resistivity measurements (Fig. 3.1). For the temperature-independent high-frequency extrapolation, $\omega \rightarrow \infty$, we utilized published reflectivity data, obtained in ultra-high vacuum for up to 170 000 cm^{-1} (21 eV) [187], and applied the procedure proposed by Tanner [145].

In order to check the robustness of our Kramers-Kronig analysis, we replaced the reflectivity data of [187] by three other data sets available in the literature for the frequency range from approximately 1–5 eV. Namely, we used: (i) the polycrystalline data from [188]; (ii) the single-crystal data from the same reference; and (iii) the most recent single-crystal data from Kozlov et al. [189]. All these data sets, as well as the data of [187], show rather similar overall reflectivity with two strong peaks at approximately 1.8 and 4 eV; the peak at 1.8 eV shows up in our data as well.

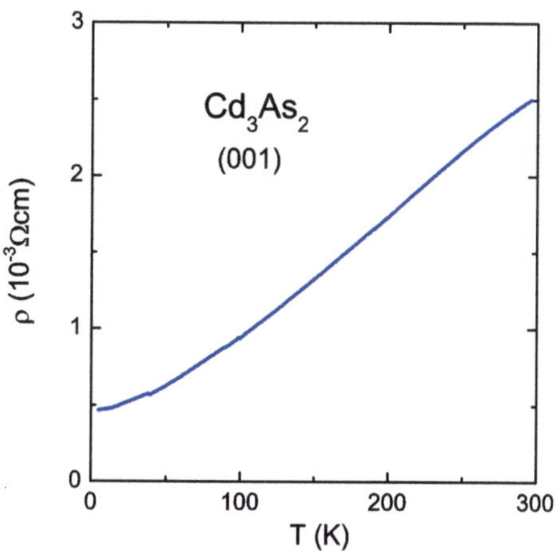

Fig. 3.1 Temperature-dependent dc resistivity of the Cd_3As_2 sample, used for optical studies, measured within the (001)-plane. Data from [172]

We have found that at frequencies below 1.2 eV (10 000 cm^{-1}) all these different extrapolations affect the outcome of the Kramers-Kronig analysis only marginally. However, it is essential not to neglect the strong peak in reflection at around 4 eV. If we ignore the 4-eV peak in our Kramers-Kronig analysis and use instead the free-electron (ω^{-4}) extrapolation right from the highest data point, the exponent of the power-law dependence of the optical conductivity between 2000 and 8000 cm^{-1} (discussed below) will be reduced from 1.65 to 1.2.

3.1.3 Experimental Results

Within the (001) plane, the optical reflectivity exhibited no dependence on the polarization. Hereafter, we present the data obtained for $\mathbf{E}_\omega \parallel [110]$ where $\mathbf{E}_\omega$ is the electric-field component of the probing light.

Figure 3.2 shows the reflectivity, $R(\omega)$, of [001]-oriented Cd_3As_2 versus frequency at various temperatures as indicated. At low frequencies the reflectivity is rather high corresponding to the metallic behavior of the dc resistivity. At $\omega_{pl}^{scr}/2\pi \approx 400$ cm^{-1} a temperature-independent (screened) plasma edge is observed in the reflectivity. A number of phonon modes strongly affect the reflectivity at lower ω; they become sharper as the temperature is reduced. For $\omega/2\pi > 2000$ cm^{-1} the reflectivity is

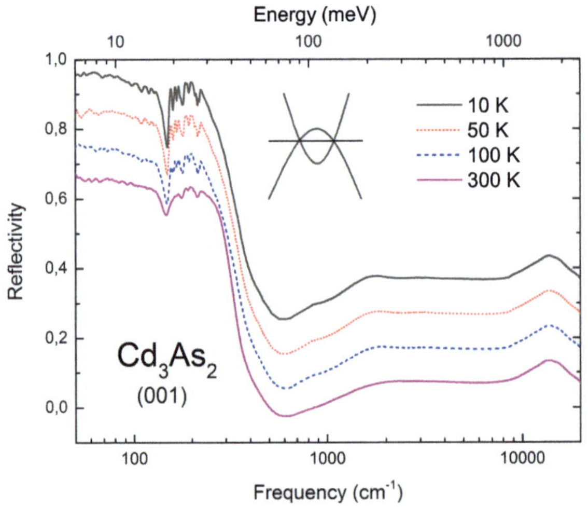

Fig. 3.2 Reflectivity of [001]-cut Cd_3As_2 for selected temperatures between 10 and 300 K measured with $\mathbf{E}_\omega \parallel [110]$, see text. Ordinate numbers are given for the measurements at 10 K, while the curves obtained at $T = 50$, 100, and 300 K are downshifted by −0.1, −0.2, and −0.3, respectively. Note the logarithmic frequency scale. The inset sketches the dispersion of the Dirac bands in Cd_3As_2. In general, the Fermi-level position is not necessarily at the Dirac points. Reproduced with permission from [172]. Copyright (2016) by the American Physical Society

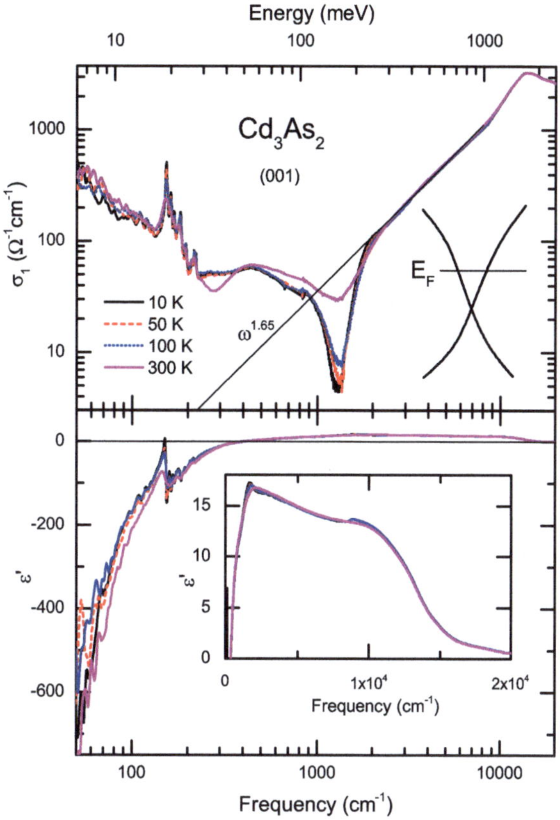

Fig. 3.3 Overall optical conductivity (upper frame) and dielectric permittivity (bottom frame) of Cd_3As_2 measured within the (001)-plane. Note the logarithmic frequency scale in the main frames and the logarithmic conductivity scale in the upper frame. The straight line in the upper frame represents $\sigma_1(\omega) \propto \omega^{1.65}$. The diagram in the upper frame sketches the proposed band dispersion in Cd_3As_2 near one of the Dirac points. The inset of the bottom frame displays positive ε' on a linear frequency scale. Reproduced with permission from [172]. Copyright (2016) by the American Physical Society

basically temperature independent. Between 2000 and 8000 cm^{-1} it is flat, followed by a broad bump centered around 14 000 cm^{-1}.

The results of the Kramers-Kronig analysis are plotted in Fig. 3.3. In this chapter, we present the data in terms of $\sigma_1(\omega)$ and the real part of the dielectric constant, $\varepsilon'(\omega)$.

The first striking result of our investigation is a power-law behavior of $\sigma_1(\omega)$ between approximately 2000 and 8000 cm^{-1}: $\sigma_1(\omega) \propto \omega^n$ with $n = 1.65 \pm 0.05$. This power-law conductivity is basically independent on temperature. As already noted, if we ignore the 4-eV peak in $R(\omega)$ in our Kramers-Kronig analysis and instead use the free-electron (ω^{-4}) extrapolation right from the highest data point, n is reduced from 1.65 to 1.2.

The straightforward application of (1.33) yields $z \simeq 0.6$, i.e. a sub-linear dispersion, $E(k) \propto |k|^{0.6}$, of the Dirac bands in Cd_3As_2. This dispersion is valid only for the energies above 250 meV (~2000 cm^{-1}), as the observed power law in $\sigma_1(\omega)$ does not extend below this frequency. Also, (1.33) does not take into account the asymmetry between the valence and conduction bands, which is present in Cd_3As_2 [12, 177]. Hence, the obtained dispersion is basically an effective approximation.

Nevertheless, the sub-linear dispersion at high energies is in qualitative agreement with the dispersion derived from Landau-level spectroscopy [177], with ARPES results [14], as well as with band-structure calculations [13, 177]. One could imagine that the Dirac bands in Cd_3As_2 get more narrow when the Dirac point is approached as depicted by the sketch in Fig. 3.3. Eventually, the bands become linear, but we cannot probe the linear dispersion by optical-conductivity measurements, because, as it will be discussed below, the Fermi level in our sample is shifted with respect to the Dirac point.

The dielectric constant (bottom panel of Fig. 3.3) is negative at low frequencies and crosses zero at 400 cm^{-1}. The crossing point is independent on temperature and is set by the screened plasma frequency of the free carriers, ω_{pl}^{scr}. Very similar values of ω_{pl}^{scr} have been reported previously [183–185] indicating similar carrier concentrations in naturally grown samples. At higher frequencies $\varepsilon'(\omega)$ takes a positive sign, reaching values up to 17 at 1500–2000 cm^{-1}.

3.1.4 Discussion

As discussed in Chap. 2, in the simplest case of symmetric 3D Dirac cones, the slope of the linear $\sigma_1(\omega)$ due to interband contributions is proportional to the (isotropic) Fermi velocity of Dirac fermions, see (1.34). If the Fermi level is not at the Dirac point ($E_F \neq 0$), (1.34) is replaced by (1.35).

In Fig. 3.4 we re-plot $\sigma_1(\omega)$ on a double-linear scale as relevant for further considerations. From Figs. 3.3 and 3.4 one can see that the low-temperature $\sigma_1(\omega)$ almost vanishes ($\sim$5 $\Omega^{-1}cm^{-1}$) at around 1300 cm^{-1} (160 meV). The power-law conductivity starts at $\sim$2000 cm^{-1} (250 meV). Following [80] and (1.35), we associate this step-like feature in $\sigma_1(\omega)$ with the position of the Fermi level.

The power-law conductivity discussed above can be roughly approximated by a straight line (the best fit is achieved between 3000 and 4500 cm^{-1}). By setting $N_W = 4$ (two spin-degenerate cones) in (1.34), we obtain $v_F = 1.7 \times 10^5$ m/s. Alternatively, one can estimate v_F from the derivative of the interband conductivity, i.e. from $\sigma_1(\omega)$ at $\nu > 2000$ cm^{-1}:

$$v_F(\omega) = \frac{e^2 N_W}{12h} \left(\frac{d\sigma_1}{d\omega} \right)^{-1}, \tag{3.1}$$

as it is shown in the inset of Fig. 3.4. As it can be seen from the Figure, the typical values of v_F (1.2×10^5 to 3×10^5 m/s) are somewhat smaller than the published results, which range from 7.6×10^5 m/s to 1.5×10^6 m/s [12–14, 177–179, 193]. Note however, that we have evaluated an energy-dependent Fermi velocity: v_F strongly increases as ω is reduced. In other words, v_F might actually meet the literature values when approaching the Dirac points. The rise of v_F at $\omega \to 0$ is exactly the behavior expected for narrowing Dirac bands.

In order to get a more quantitative description of the optical conductivity in Cd_3As_2, one needs to implement a model. Below, we will use a simple model of

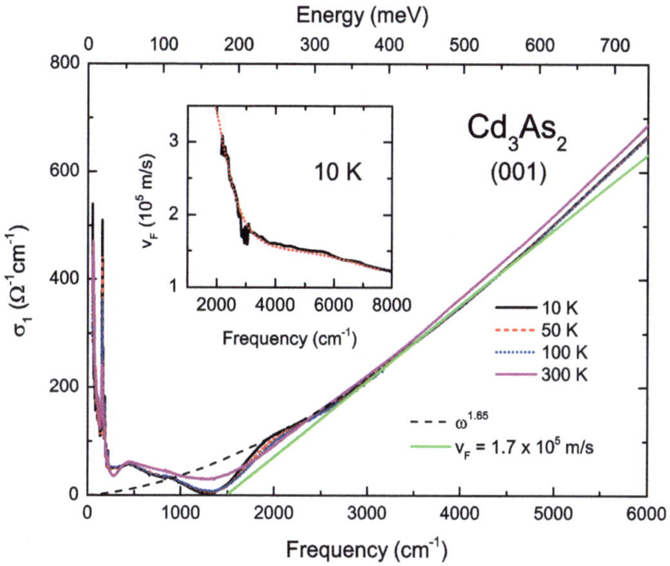

Fig. 3.4 Optical conductivity from Fig. 3.3 re-plotted on double-linear scale for $\omega < 6000\ \text{cm}^{-1}$. The straight green line represents the linear conductivity with $v_F = 1.7 \times 10^5$ m/s. The inset shows the Fermi velocity calculated using (3.1) from $\sigma_1(\omega)$ at 10 K. Red dotted line is a guide to the eye. Reproduced with permission from [172]. Copyright (2016) by the American Physical Society

linear $E(k)$ spectrum with the purpose of extracting the Fermi-level position and explaining its broadening. This model will also effectively describe the leveling off the interband conductivity spectrum at low frequencies. A more elaborative model for the optical conductivity of Cd_3As_2 should also take into account the electron-hole asymmetry of the Dirac bands as well as the presence on the Lifshitz transition. Building such a model goes beyond the present work. Let us note that the position of the Lifshitz-transition point is debatable. Calculations performed by Feng et al. predict the Lifshitz transition to be situated at 133 meV aside the Dirac points (see supplementary material to [194]). According to Jeon et al. [177], however, the distance between the Lifshitz-transition and Dirac points is rather in the 10-meV range.

Our model is based on considering self-energy effects in the spirit of [195] and is provided in [172]. The real parts of the conduction and valence-band self energies are approximated by the constant $\text{Re}\Sigma_c = \Delta$ and $\text{Re}\Sigma_v = -\Delta$, respectively. We hence obtain for the interband conductivity

$$\sigma_1(\omega) = \frac{e^2 N_W}{12 h v_F} \frac{(\omega - \omega_g)^2}{\omega} \theta \left\{\hbar\omega - 2\max[E_F, \Delta]\right\}, \tag{3.2}$$

where $\hbar\omega_g = 2\Delta$. The conduction-band energies have been pushed up by Δ, while the valence-band energies are lowered by the same amount. Alternatively, (3.2) is the

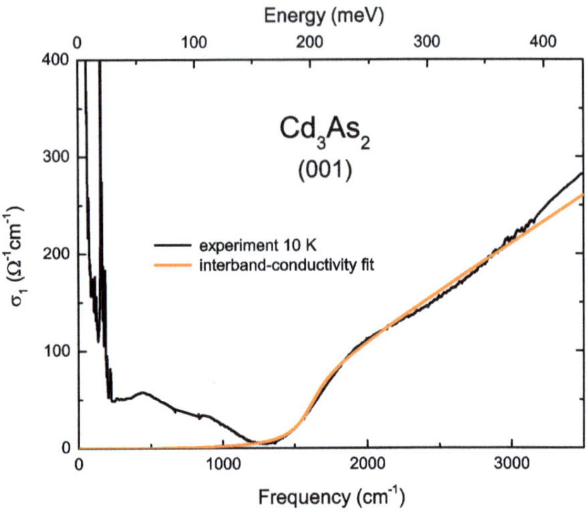

Fig. 3.5 Optical conductivity of Cd_3As_2 measured within the (001) plane at $T = 10$ K, plotted together with a description of the interband portion of $\sigma_1(\omega)$ given by (3.2) and (3.3). Reproduced with permission from [172]. Copyright (2016) by the American Physical Society

interband conductivity of a simple band structure with $E(k) = \pm[\hbar v_F k + \Delta]$, which could be thought of as a first rough approximation to the case of the Dirac cones that get narrow as energy approaches the Dirac point. The Fermi energy is measured from the Dirac point without self-energy corrections included, and so E_F must be larger than Δ for finite doping away from charge neutrality. If impurity scattering cannot be neglected, the Heaviside function can be replaced by

$$\frac{1}{2} + \frac{1}{\pi} \arctan \frac{\omega - 2\max[E_F, \Delta]/\hbar}{\gamma}. \tag{3.3}$$

where γ represents a frequency-independent impurity scattering rate.[1]

For $\Delta > E_F$, (3.2) is reduced to the form proposed by Morimoto and Nagaosa for the optical conductivity in so-called Weyl-Mott insulators [196]. At $\omega \gg \omega_g$, (3.2) becomes linear in frequency, similarly to (1.34) and (1.35). However, differently from (1.35), the interband conductivity approaches zero at $\omega = \omega_g$, rather than at $\omega = 0$, in the case of $E_F = 0$.

A combination of (3.2) and (3.3) is now employed to model the experimental data; the best fit is plotted in Fig. 3.5. Note, that this model does not include the intraband conductivity. The best description of the experimental curve was achieved with $E_F/hc = 800\,\mathrm{cm}^{-1}$ ($E_F \approx 100\,\mathrm{meV}$), $\omega_g/2\pi = 450\,\mathrm{cm}^{-1}$, $\gamma/2\pi = 120\,\mathrm{cm}^{-1}$, and $v_F = 2.4 \times 10^5$ m/s. The obtained value of ω_g, which is found to be smaller than the Fermi energy, has to be considered as a fit parameter only. The fit is not perfect, as the model doesn't include the deviations from the band linearity discussed above. Nevertheless, the model grasps the main features of the interband conductivity in Cd_3As_2.

[1] For convenience, we sometimes use slightly different notations in different chapters, thus (1.36), (2.2), and (3.3) have slightly different forms.

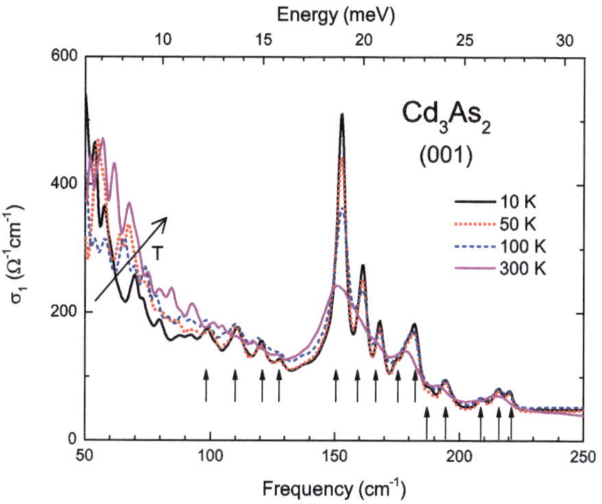

Fig. 3.6 Low-frequency part of optical conductivity in Cd_3As_2. The vertical arrows indicate the positions of the phonons. Reproduced with permission from [172]. Copyright (2016) by the American Physical Society

The obtained position of the Fermi level (100 meV) seems to be quite reasonable for our sample, taking into account its carrier concentration (6×10^{17} cm^{-3}) and keeping in mind that $E_F = 200$ meV was reported for a sample with $n_e = 2 \times 10^{18}$ cm^{-3} [177] and $E_F = 286$ meV for $n_e = 1.67 \times 10^{18}$ cm^{-3} [193].

Although here we mostly concentrate on the interband conductivity in Cd_3As_2, let us briefly discuss the experimental results at the lowest frequencies measured. The intraband conductivity is represented by a narrow Drude component (best seen in Fig. 3.6) and an absorption band of peculiar shape at 300–1300 cm^{-1} (see Figs. 3.4 and 3.5). The nature of the band might be related to localization effects. The presence of this band makes it impossible to fit the intraband conductivity for $\nu < 1300$ cm^{-1} with a simple Drude term. One can see, however, that the narrow Drude peak is getting somewhat narrower as $T \rightarrow 0$ due to a modest decrease of scattering. Let us note that the spectral weight related to the low-frequency absorption of delocalized carriers remains temperature independent, because the screened plasma frequency, discussed above in connection to the dielectric constant, is independent of temperature.

In addition to electronic contributions, the low-frequency conductivity renders a large number of phonon modes marked by arrows in Fig. 3.6. We can distinguish 14 infrared-active phonon modes at the frequencies between approximately 100 and 250 cm^{-1}. As the temperature is reduced, the phonons become sharper.

3.1.5 Conclusions

We conclude that the dc resistivity and the optical conductivity of [001]-oriented n-doped ($n_e = 6 \times 10^{17}$ cm^{-3}) Cd_3As_2 are isotropic within the (001)-plane. The real part of the frequency-dependent conductivity follows a power law, $\sigma_1(\omega) \propto \omega^{1.65}$, in a broad frequency range, 2000–8000 cm^{-1}. We interpret this behavior as the manifestation of interband transitions between two Dirac bands with a sub-linear dispersion relation, $E(k) \propto |k|^{0.6}$. The Fermi velocity falls in the range between 1.2×10^5 and 3×10^5 m/s, depending on the distance from the Dirac points.

At 1300 cm^{-1} (160 meV), we find a diminishing conductivity, consistent with observations of an "optical gap", made in the 1960s [181]. However, we hesitate to follow the traditional interpretation of this feature and straightforwardly relate it to a gap in the density of states. Applying recent models for the optical response of Dirac/Weyl semimetals, we instead relate this feature to the Fermi level, which is positioned around 100 meV above the Dirac points, and which is consistent with the carrier concentration.

3.2 The Dirac Semimetal Au_2Pb

Recently, Au_2Pb gained attention because of the prediction [197] of Dirac bands in this Laves-phase compound, which was known to be superconducting ($T_c \simeq 1.2$ K) for decades [198]. Au_2Pb possesses three structural transitions at $T = 97$, 51, and 40 K [197, 199–202]. The high-temperature structure is cubic (space group $Fd\bar{3}m$) and the electronic structure features bulk Dirac cones (six per BZ), as confirmed by ARPES [202]. The Dirac points are protected by C_4 rotation symmetry. Upon cooling, the symmetry gets lowered and the Dirac bands become gapped. At the lowest temperatures, the structure is orthorhombic (space group $Pbcn$). This phase has also been predicted to be topologically nontrivial and to possess topological surface states with linear dispersion relations [197]. In the superconducting state, the surfaces of Au_2Pb crystals are therefore a natural platform for realizing Majorana fermions [203]. Most recent angle-resolved photoemission spectroscopy (ARPES) provided an experimental evidence for Dirac states in Au_2Pb [202].

A direct gap is present in the low-temperature electronic structure at all momenta, but a number of electron- and hole-like trivial bands cross the Fermi level. The trivial electron- and hole pockets also exist in the high-temperature phases, in addition to the bulk Dirac bands. Carriers from these trivial Fermi-surface pockets dominate the broad optical response of Au_2Pb. Following our work [173], we investigate their contributions to the optical conductivity in this chapter.

3.2.1 Sample Preparation and Characterization

Single crystals of Au_2Pb were synthesized according the procedure reported by Schoop et al. [197]. The single crystals were grow out of lead flux. For the synthesis, 0.78 g of gold powder (99.99%) and 1.24 g of lead beads (99.999%, average diameter is 1 mm) were put together in a quartz tube with a neck of about 1–2 mm in the middle of the tube. The evacuated and sealed tube was heated within 12 h to 600 °C and kept there for 24 h. Then the melt was allowed to cool down

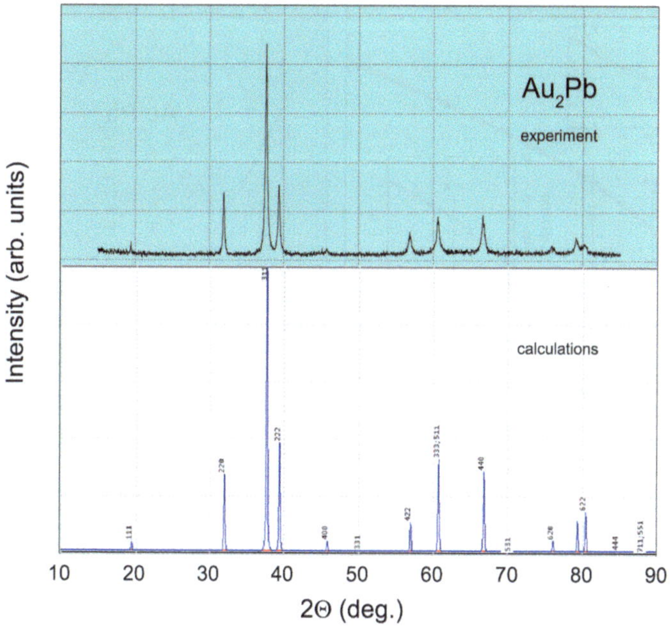

Fig. 3.7 X-ray diffraction pattern of milled Au_2Pb powder (top panel) in comparison with calculations (bottom panel). Data from [173]

(3 °C/h) to 300 °C and kept there for 48 h. When the quartz glass tube was turned, flux and crystals were separated. The traces of remaining lead on the surface of the crystals were removed by washing them in an aqueous solution of 20 ml acetic acid (50%) and 4 ml hydrogen peroxide (30%) for some minutes. Powder x-ray diffraction data of ball milled samples were collected on a Bruker D2 Phaser using MoKα-radiation ($\lambda = 0.71073$ Å) at room temperature and indicate a cubic Laves phase (space group $Fd\bar{3}m$) with the lattice constant $a = 7.911(1)$ Å. No reflections from impurities, elements (Au and Pb), and any other lead-gold phases have been detected in these measurements, see Fig. 3.7. For the crystals used in the measurements, Laue diffraction demonstrated perfect single-crystalline patterns with the largest surfaces being either (111) or (100).

Electrical dc resistivity measurements were performed in four-contact geometry in a custom-made setup by cooling from room temperature down to 5 K. Transversal magnetoresistance and Hall resistivity measurements were conducted on Hall bars at 5 to 280 K in magnetic fields B of up to 6 T. The crystal plane used for all transport experiments was (110). We should notice that due to the cubic high-temperature structure and the fact the at low temperatures the orthorhombic distortions are pretty small, no significant anisotropy is expected in Au_2Pb [197].

The results of our dc measurements are shown in the panel (a) of Fig. 3.8. A clear metallic behavior is observed, the residual resistivity being $\rho_0 = 5.5\ \mu\Omega$cm and $\rho(300\ \text{K})/\rho_0 = 7.5$. These values, as well as the overall resistivity curve, are in

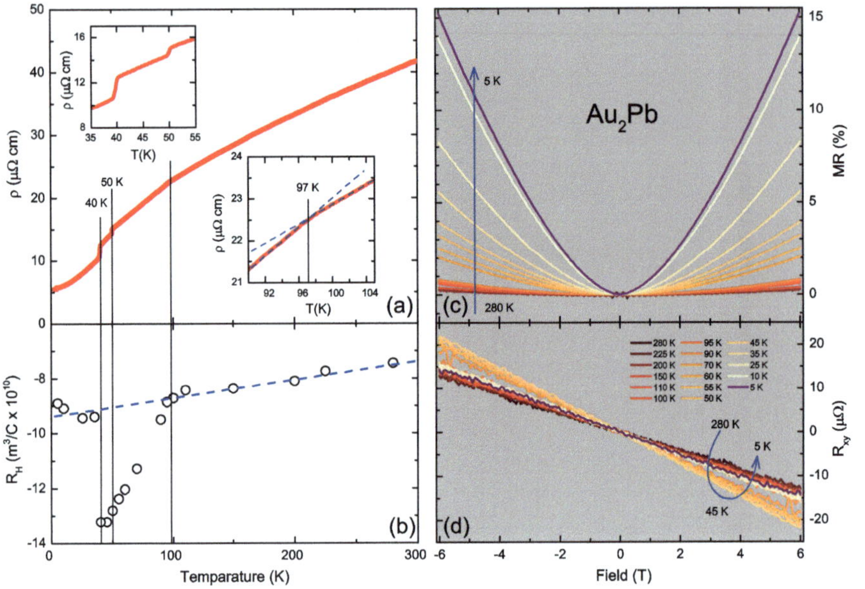

Fig. 3.8 Panel **a**: Electrical dc resistivity of Au_2Pb as a function of temperature. The insets zoom in the areas near the structural transitions at 40, ~50, and 97 K. Panel **b**: Hall coefficient as a function of temperature (open dots). The structural transitions are indicted as thin vertical lines in the panels (**a**) and (**b**). In both panels, dashed blue lines serve as guides to the eye. Panel **c**: Transversal magnetoresistance (in percent) as a function of applied magnetic field for a number of temperatures between 5 and 280 K. Panel **d**: Hall resistance as a function of magnetic field for the temperatures indicated.

good agreement with results previously reported in literature [197, 199, 200]. As discussed above, Au_2Pb undergoes several structural phase transitions with decreasing temperature: at 97, 51, and 40 K [197]; accordingly, anomalies in the measured $\rho(T)$ are detected at these temperature. The magnetotransport results are presented in the panels (b), (c), and (d). We observe a non-saturating moderate positive magnetoresistance at all measurement temperatures and magnetic fields, panel (c). The Hall constant is negative and non-monotonic in temperature [panels (b) and (d)]. It exhibits jumps at the structural-transition temperatures, reflecting the corresponding changes in band structure [197].

As already noticed, band-structure calculations for the high- and low-temperature phases predict that both, electron- and hole-like, bands cross the Fermi level and, hence, contribute (with different signs) to R_H at any temperature. Thus, a straightforward determination of the free-carrier concentration from R_H is not possible. Application of the one-carrier-type (OCT) relation, $R_H = 1/(en)$, allows a low-bound estimation for the free-carrier concentration, $n \sim 5 \times 10^{21}$ cm^{-3} at 5 K. Such high n strongly indicates conduction of metallic type (obviously, the real carrier concentrations can only be higher, as holes and electrons provide Hall constants of opposite signs, which partly compensate each other). Within the OCT model, one can also obtain rough estimates of the carrier mobility, $\mu^{OCT}(5\text{K}) \sim 10^2$ cm^2/(Vs), and of the relaxation rate, $\gamma^{OCT}(5\text{K}) \sim 100$ cm^{-1} (here, the free-electron mass is taken as the effective carrier mass). These values are obviously some sort of averages over all the bands crossing the Fermi level and hence should be considered with a grain of salt. Still, γ^{OCT} lies right in-between of the scattering rates obtained from our optical studies for the carriers described by the so-called narrow and broad Drude components, as discussed below (we use indices N and B hereafter to mark the quantities related to these two components, respectively).

3.2.2 *Optical Experiments*

Temperature-dependent optical reflectivity, $R(\nu)$, was measured on a large [3 × 2 mm^2] polished (100) surface of a Au_2Pb single crystal over a broad frequency range from 30 to 10 000 cm^{-1}. Additionally, we measured room-temperature reflectivity up to 48 000 cm^{-1}. The spectra in the far-infrared (below 700 cm^{-1}) were collected with a Bruker IFS 113v Fourier-transform spectrometer using in situ gold coating of the sample surface for reference measurements. At higher frequencies (up to 22 000 cm^{-1}), a Bruker Hyperion infrared microscope attached to a Bruker Vertex 80v FTIR spectrometer was used. For these measurements, freshly evaporated gold mirrors (below 10 000 cm^{-1}) and protected silver (above 10 000 cm^{-1}) served as reference. Finally, at the highest frequencies (up 48 000 cm^{-1}) we used a Woollam variable-angle spectroscopic ellipsometer with SiO_2 on Si substrate as reference.

For Kramers-Kronig analysis, zero-frequency extrapolations have been made using the Drude model in accordance with the temperature-dependent dc resistivity measurements. Two narrow gaps in the measurements at around 100 and 700 cm^{-1}, originating from absorption in beam splitters, were bridged by linear interpolations. For high-frequency extrapolations, we utilized the x-ray atomic scattering functions [145] followed by the free-electron behavior, $R(\omega) \propto 1/\omega^4$, above 30 keV.

3.2.3 *Results and Discussion*

Figure 3.9 displays the optical reflectivity $R(\nu)$, the real part of the optical conductivity $\sigma_1(\nu)$ and of the dielectric constant $\varepsilon_1(\nu)$, as well as the skin depth $\delta(\nu)$ of Au_2Pb for different temperatures. For frequencies higher than ~10 000 cm^{-1}, the optical properties are independent on temperature. Let us immediately note that the skin depth exceeds 20 nm for all measured temperatures and frequencies. Hence, our optical measurements of Au_2Pb reflect its bulk properties.

Au_2Pb demonstrates a typical metallic response: at $\nu \to 0$ the reflectivity approaches unity, $\varepsilon(\nu)$ is large and negative, $\sigma_1(\nu)$ is very high, and $\delta(\nu)$ is inversely

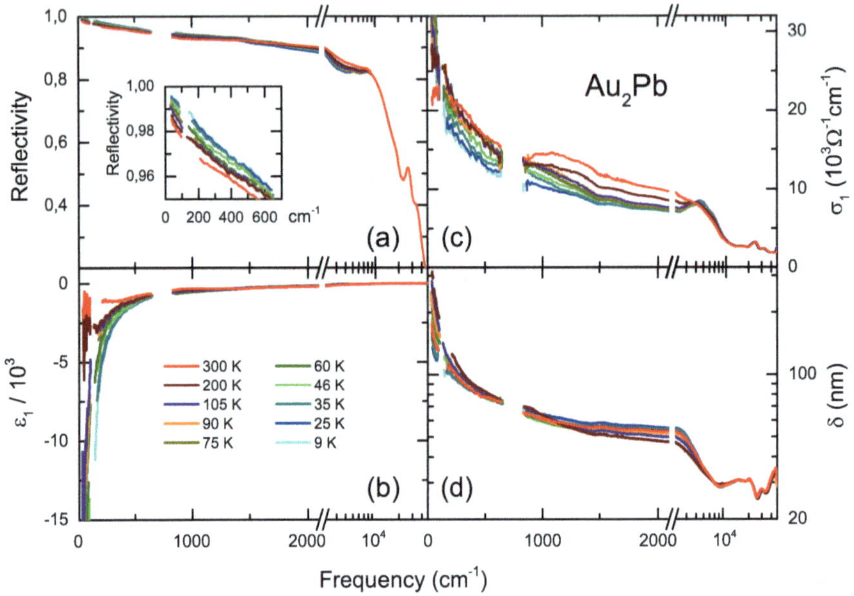

Fig. 3.9 Optical reflectivity (**a**), real parts of the dielectric permittivity (**b**) and optical conductivity (**c**), and skin depth (**d**) of Au_2Pb at selected temperatures between $T = 9$ and 300 K; note the x-scale change at 2000 cm^{-1}. The inset shows the low-frequency reflectivity on an enlarged scale. Data are adopted from [173]

proportional to frequency. $R(\nu)$ starts to drop down at around 10 000 cm^{-1}, indicating the onset of the plasma edge. This high-frequency position of the plasma frequency is another indication of the high free-carrier concentration in Au_2Pb.

Further, we can conclude that high concentration of free carriers prevents us from observing any sort of characteristic optical features of a Dirac material, where the interband transitions between the linearly dispersing bands are supposed to manifest themselves in $\sigma_1(\nu)$ according to (1.33).

In Au_2Pb, one could expect to see the optical transitions between the lower and the upper Dirac bands at frequencies below some 3 200 cm^{-1}, as the Lifshitz transition is supposed to be at around 400 meV $\cong$ 3 200 cm^{-1} [197]. To estimate the expected interband optical conductivity due to the inter-Dirac-band transitions σ_1^{Dirac}, we can utilize (1.34), which connects the real part of the complex conductivity to the number of non-degenerate bands per BZ N_W and the Dirac-band Fermi velocity v_F. Using $N_W = 2N_D = 12$ (N_D is the number of Dirac cones per BZ) and $v_F \sim 10^6$ m/s, estimated from the band-structure calculations [197], we calculate the expected optical conductivity due to the transitions between the lower and the upper Dirac bands. We find this conductivity to be orders of magnitude below the measured values. For example, we obtain σ_1^{Dirac} of the order of 100 $\Omega^{-1}\text{cm}^{-1}$ at 1000 cm^{-1} (cf. the experimentally observed value, $\sigma_1 \sim 10^4\ \Omega^{-1}\text{cm}^{-1}$). Hence, the inter-Dirac-band transitions are completely masked in Au_2Pb by free carriers. Hereafter, we concentrate on the intra-band, i.e. free-carrier, optical response.

In order to fit the optical response of Au_2Pb, we used a standard Drude-Lorentz ansatz [75]. We have found that best description can be obtained, if we utilize two Drude contributions (narrow and broad, N and B, as mentioned above). Being expressed in terms of complex optical conductivity, our fitting function reads as:

$$\sigma = \frac{\sigma_0^N}{1 - i\omega\tau^N} + \frac{\sigma_0^B}{1 - i\omega\tau^B} + \frac{\omega}{4\pi}\sum \frac{\Delta\varepsilon_k \omega_{0k}^2}{i(\omega_{0k}^2 - \omega^2 - i\omega\Gamma_k)}. \tag{3.4}$$

Here, σ_0 is the dc limit of a Drude term, τ represents a free-electron scattering time, ω_{0k} is the eigenfrequency of k-th Lorentz oscillator, $\Delta\varepsilon_k$ is its dielectric contribution and Γ_k is the corresponding line width.

As already mentioned, band-structure calculations and ARPES measurements demonstrate that at any temperature Au_2Pb possesses a few bands (Dirac or not) crossing the Fermi level [197, 202]. Thus, a multi-component Drude fit is relevant. Numerous previous optical-conductivity studies show that two Drude components is typically sufficient to fit the spectra, even if the system under study possesses more than two bands: first, adding further Drude terms (>2) leads to ambiguous fits; second, a given Drude term is not necessarily related to the scattering processes within a given conduction band. Thus, we stick to the common minimal model. Such two-component approach has been widely used to interpret the normal-state optical spectra of iron-based superconductors [204, 205]. It has also been applied to various topological semimetals, as discussed in the following chapters. In the case of topological semimetals, the two Drude terms can sometimes be interpreted as the response of free carriers within the Dirac (highly mobile carriers) and parabolic (less mobile carriers) bands. We show below that this simple interpretation is not applicable to Au_2Pb.

Figure 3.10a, b shows the results of fitting the experimental data with (3.4) at 9 and 300 K. Let us note, that for best possible model description of the spectra, we fitted the experimental spectra of $R(\nu)$, $\sigma(\nu)$ and $\varepsilon(\nu)$ simultaneously. Also, we did not impose any restrictions on the fit parameters. In addition to the Drude terms, we used five Lorentzians, which effectively describe multiple transitions between trivial bands at high frequencies. We found that the two-Drude approach is relevant at all measurement temperatures, except of 300 K, where the two Drude components are not distinguishable and the spectra can be described by a single Drude term.

The fit parameters of the Drude terms versus temperature are shown in Fig. 3.10c, d. As one can see from the panel (c), the two Drude components completely account for the dc conductivity, $\sigma_{dc}(T) \equiv 1/\rho(T)$, of Au_2Pb: the sum of the zero-frequency limits of the Drude modes matches the dc conductivity values perfectly at any temperature. It is also evident that it is the narrow component, which is mostly responsible for the temperature dependence of σ_{dc}. Overall, the temperature evolution of the optical spectra is pretty smooth.

We have to notice that the broad Drude term does reflect the behavior of free carriers, has nothing to do with localization, and cannot be replaced by a Lorentzian. As one can see from Fig. 5.2d, this term has an almost temperature-independent

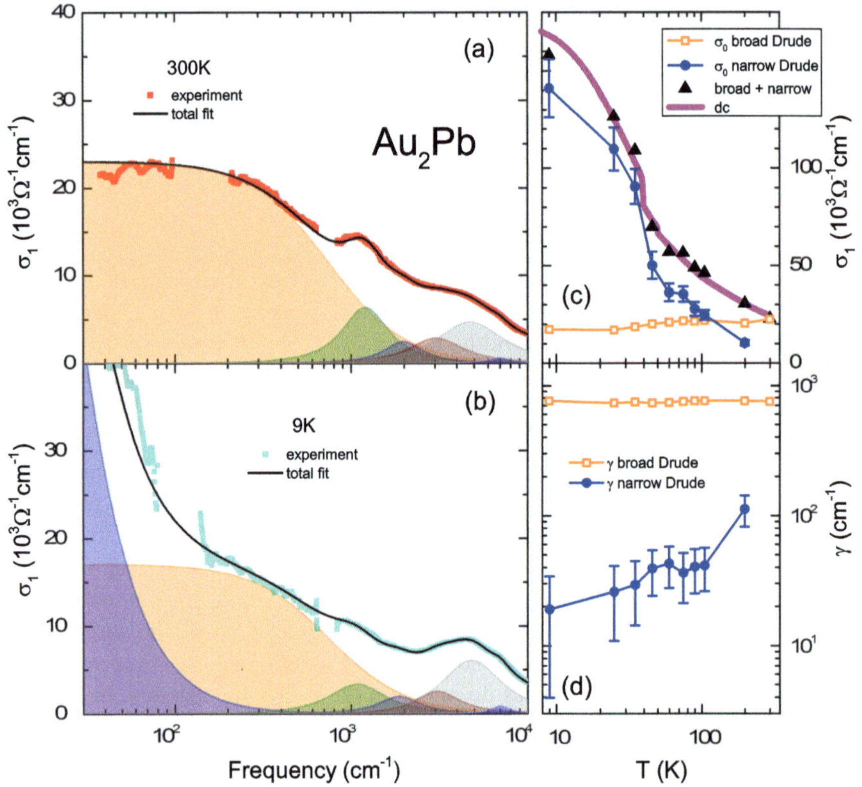

Fig. 3.10 Drude-Lorentz fits of the optical conductivity of Au_2Pb at 300 K (**a**) and 9 K (**b**). Shadowed areas of different colors correspond to Lorentzians and the two Drude terms (broad and narrow) as discussed in the text. Parameters of the Drude terms as functions of temperature: the zero-frequency limit σ_0 (**c**) and the scattering rate γ (**d**). Typical fit error bars for σ_0 and γ of the narrow Drude term are displayed; the error bars for the broad Drude term are within the symbols. Bold magenta line in panel (**c**) shows the dc conductivity.

scattering rate, $\gamma^B = 1/(2\pi\tau^B) = 750$ cm^{-1}. An averaged Fermi velocity of the non-Dirac bands in Au_2Pb can be estimated from the band structure [197] and is in the 10^5 m/s range (for Dirac bands, it is even higher, as discussed above). Thus, the mean free path of the carriers, responsible for the broad Drude term, is at least around 7 Å. This is a few times more than the distance between the atoms. Hence, unlike the situation, e.g., in the superconducting cuprates [206, 207] and some other materials [208–210], the Ioffe-Regel limit [211] is not violated in Au_2Pb. It is not a bad metal [212], the transport in Au_2Pb is coherent and our "broad Drude" description is relevant.

The plasma frequencies of the two Drude components, ν_{pl}^B and ν_{pl}^N, are shown in Fig. 3.11. They have generally rather flat temperature dependencies.

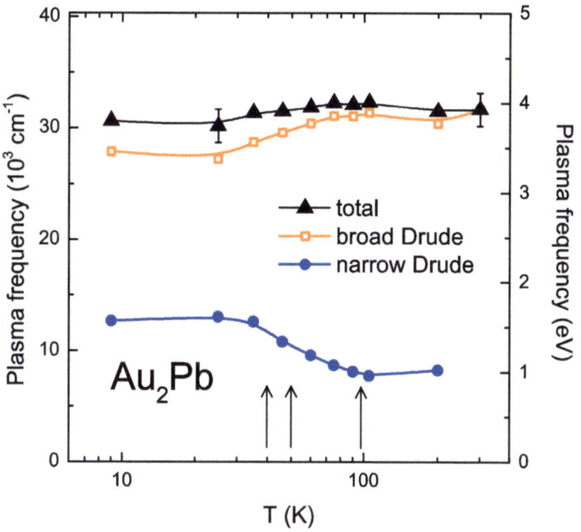

Fig. 3.11 Plasma frequencies of the broad and narrow Drude components and total plasma frequency of free carriers in Au_2Pb as functions of temperature. The structural transitions (cf. Fig. 3.8) are marked with arrows. ©IOP Publishing. Reproduced with permission from [173]. All rights reserved

Nevertheless, the plasma frequency (and hence, the corresponding spectral weight, which is proportional to the carrier density) of the narrow Drude band increases upon cooling. Would this absorption band correspond to the bulk Dirac carriers, $\nu_{pl}^{N}(T)$ had to decrease instead and eventually to disappear as $T \to 0$, because the bulk Dirac bands are gapped and do not cross the Fermi level anymore in the low-temperature phase of Au_2Pb. As this does not happen, one can conclude that narrow Drude term is not directly related to Dirac electrons. Obviously, the bulk Dirac electrons cannot provide any significant contribution to the broad Drude term, as this term is present at all temperatures and its spectral weight has hardly any temperature dependence. Also, Dirac fermions are typically highly mobile [17, 179], their scattering rates are low and the corresponding Drude bands are very narrow, see, e.g., the previous chapter and the sections on Cd_3As_2 and NbP. Overall, the contribution of free Dirac carriers to the total optical response is basically negligible and optics only probes the carriers in the trivial bands of Au_2Pb for all its structural phases.

In Fig. 3.11, the temperatures of the structural transitions are marked with vertical arrows. It is apparent that at temperatures around these transitions, the spectral weight redistributes between the Drude components, reflecting changes in the band structure. One has to note that plasma frequency is determinate by both, free-carrier concentration and effective mass. Thus, sharp changes in the Hall coefficient need not be directly reflected in the temperature evolution of plasma frequency. The smooth temperature evolution of ν_{pl}^{B} and ν_{pl}^{N} between roughly 40 and 100 K (at higher and lower temperatures, they are temperature independent within our accuracy) might indicate that more than one crystallographic structure exists at a given temperature here. Such nonhomogeneous states have indeed been detected in Au_2Pb by temperature-dependent x-ray diffraction at a number of temperatures in this intermediate-temperature range [197].

As one can see from Fig. 3.11, the total plasma frequency is almost temperature independent, in agreement with the metallic type of conduction, see Fig. 3.8. The temperature-averaged value of the total plasma frequency is $\nu_{pl} = \sqrt{(\nu_{pl}^B)^2 + (\nu_{pl}^N)^2} = (31\,500 \pm 1\,500)$ cm^{-1}, $h\nu_{pl} \cong (3.9 \pm 0.2)$ eV. From ν_{pl}, the n/m^* ratio can be directly computed via:

$$\omega_{pl}^2 = (2\pi\nu_{pl})^2 = \frac{4\pi e^2 n}{m^*}. \tag{3.5}$$

Here m^* is an averaged effective mass of all carriers and a parabolic band dispersion is assumed (as discussed above, we can ignore Dirac electrons, for which the plasma frequency is supposed to be related to the Fermi velocity, the band filling and, generally, the band gap [213, 214]). Thus, we obtain from (3.5): $\frac{n}{m^*/m_e} = 1.1 \times 10^{22}$cm^{-3}, where m_e is the free-electron mass. The average effective mass is not know for Au_2Pb. If we take $m^* = m_e$, we get $n \sim 10^{22}$ cm^{-3}, a reasonable metallic concentration.

3.2.4 *Conclusions*

We have measured broadband optical response of Au_2Pb—a Dirac material, which possess bulk Dirac bands at high temperatures (above 97 K) and surface Dirac bands at low temperatures (below 40 K). Neither of the Dirac states could be seen in optical conductivity because of the very high absorption due to free carriers (electrons and holes) in the trivial bulk bands. The total plasma frequency of these carriers remains temperature independent at 3.9 ± 0.2 eV. Optical measurements provide an estimate for the total carrier concentration: $\frac{n}{m^*/m_e} \simeq 10^{22}$ cm^{-3}.

3.3 The Weyl Semimetal NbP

3.3.1 *Introduction*

NbP is a nonmagnetic non-centrosymmetric Weyl semimetal from the TaAs family with extremely large magnetoresistance and ultrahigh carrier mobility [17, 21]. These extraordinary transport properties are believed to be caused by quasiparticles in linearly dispersing Weyl bands [16, 20]. According to band-structure calculations NbP possesses 24 Weyl nodes, i.e. twelve pairs of the nodes with opposite chiralities [16, 18, 20, 215, 216]. The nodes are "leftovers" of nodal rings, which are gapped by SOC everywhere in the BZ, except of these special points [16, 18, 216, 217]. The nodes can be divided in two groups, commonly dubbed as W1 ($N_{W1} = 8$) and W2 ($N_{W2} = 16$); here $N_{W1,2}$ is the number of nodes of each type. Most recent band-structure calculations agree well on the energy position of the W1 nodes: 56–57 meV below the Fermi level E_F [218–220]; the position of the W2 nodes is specified less accurately, ranging from 5 [218, 219] to 26 meV [220] above E_F.

Similar to all Weyl semimetals, physical properties of NbP are mostly determined by the low-energy electron dynamics [23, 40, 41]. Infrared optical methods enable direct access to this dynamics, as discussed above for ZrSiS, Cd_3As_2, and Au_2Pb.

In this section, we present the measurements and analysis of both, interband and itinerant-carrier, conductivity of NbP, as previously reported in [174]. We show that the low-energy interband conductivity of NbP is dominated by transitions between the bands with parallel dispersions that are split by SOC. These excitations, as well as the Drude response of the itinerant carriers, completely mask the linear-in-frequency $\sigma_1(\omega)$ due to the three-dimensional chiral Weyl bands. At somewhat higher frequencies (1400–2000 cm^{-1}, 175–250 meV), $\sigma_1(\omega)$ becomes roughly linear. Our calculations demonstrate that this linearity stems from the fact that all electronic bands, which are involved in the transitions with relevant energies, are roughly linear. In addition, we find that at low temperatures the itinerant carriers in one of the conduction channels possess a fairly long momentum-relaxing scattering time and mesoscopic characteristic length scales of momentum relaxation.

3.3.2 Sample Preparation, Experimental and Computational Details

Single crystals of NbP were synthesized according to the description reported in [17, 221]: a polycrystalline NbP powder was synthesized in a direct reaction of pure niobium and red phosphorus; the single NbP crystals were grown from the powder via vapor-transport reaction with iodine.

The electrical resistivity, $\rho_{dc}(T)$, was measured as a function of temperature within the (001)-plane in four-contact geometry. The experiments were performed on a small piece, cut from the specimen used for the optical investigations. The results of the dc measurements are plotted in Fig. 3.12. A clear metallic behavior with linear-in-temperature resistivity is observed down to approximately 100 K. Below, $\rho_{dc}(T)$ levels off and approaches the residual resistivity $\rho_0 = 0.55\ \mu\Omega$cm; the resistivity ratio is $\rho(300\ \text{K})/\rho_0 = 40$. These values are well comparable with the ones reported in literature [17, 21, 222]. Note, NbP exhibits an extremely high mobility independent on the residual resistivity ratio [21].

The normal-incidence optical reflectivity $R(\nu, T)$ was measured on the (001)-surfaces of a large (2 mm × 2 mm in lateral dimensions) single crystal from room temperature down to $T = 10$ K covering a wide frequency range from $\nu = 50$ to 12 000 cm^{-1}. The temperature-dependent experiments were supplemented by room-temperature reflectivity measurements up to 25 000 cm^{-1}. In the far-infrared spectral range below 700 cm^{-1}, a Bruker IFS 113v Fourier-transform spectrometer was employed with *in situ* gold coating of the sample surface for reference measurements.

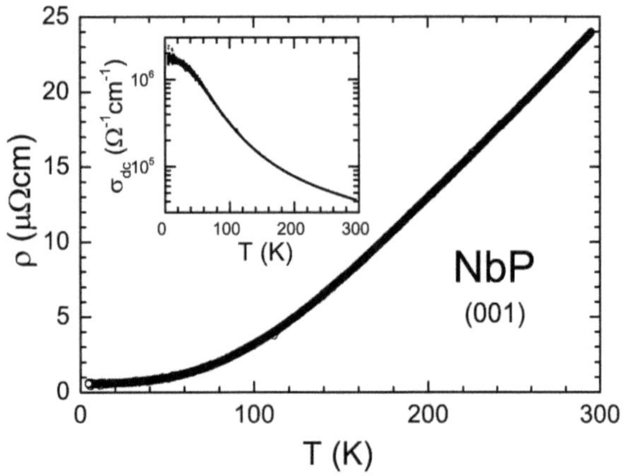

Fig. 3.12 Temperature-dependent (001)-plane dc resistivity of the NbP sample used in the optical measurements. The inset shows dc conductivity. Reproduced with permission from [174]. Copyright (2018) by the American Physical Society

At higher frequencies, we used a Bruker Hyperion infrared microscope attached to a Bruker Vertex 80v spectrometer. Here, freshly evaporated gold mirrors (below 12 000 cm^{-1}) and protected silver (above 12 000 cm^{-1}) served as reference.

For the Kramers-Kronig analysis [75] we involved the x-ray atomic scattering functions for high-frequency extrapolations [145]. From recent optical investigations of materials with highly mobile carriers, it is known [143, 146] that the commonly applied Hagen-Rubens extrapolation to zero frequency is not adequate: the very narrow zero-frequency component present in the spectra corresponds to a scattering rate comparable to (or even below) our lowest measurement frequency, $\nu_{\min} \approx 50$ cm^{-1} (see also discussion in the next section). Thus, we first fitted the spectra with a set of Lorentzians (similar fitting procedures can be utilized as a substitute of the Kramers-Kronig analysis [223–225]) and then we used the results of these fits between $\nu = 0$ and $\nu_{\min}$ as zero-frequency extrapolations for subsequent Kramers-Kronig transformations. We note that our optical measurements probe the bulk material properties, as the penetration depth exceeds 20 nm for any measurement frequency.

We performed band structure calculations within the local density approximation (LDA) based on the crystal structure of NbP determined by experiments [226]; we employed the linear muffin-tin orbital method [228] as implemented in the relativistic PY LMTO computer code. Some details of the implementation can be found in [227]. The Perdew-Wang parametrization [229] was used for the exchange-correlation potential. SOC was added to the LMTO Hamiltonian in the variational step. BZ integrations were done using the improved tetrahedron method [230]. Dipole matrix elements for interband optical transitions were calculated on a $96 \times 96 \times 96$ k-mesh using LMTO wave functions (as was shown in [231], it is necessary to use sufficiently dense meshes in order to resolve transitions between the SOC-split bands). The real part of the optical conductivity was calculated by the tetrahedron method.

3.3.3 Results and Analysis

Figure 3.13 displays the overall reflectivity $R(\nu)$, the real part of the conductivity $\sigma(\nu)$, and the dielectric constant $\varepsilon(\nu)$ of NbP for different temperatures. For frequencies higher than 5000 cm^{-1}, the optical properties are basically independent on temperature. In the spectra we can identify the signatures of: (i) phonons, (ii) itinerant-carrier (intraband) absorption, and (iii) interband transitions. Below we discus all these spectral features.

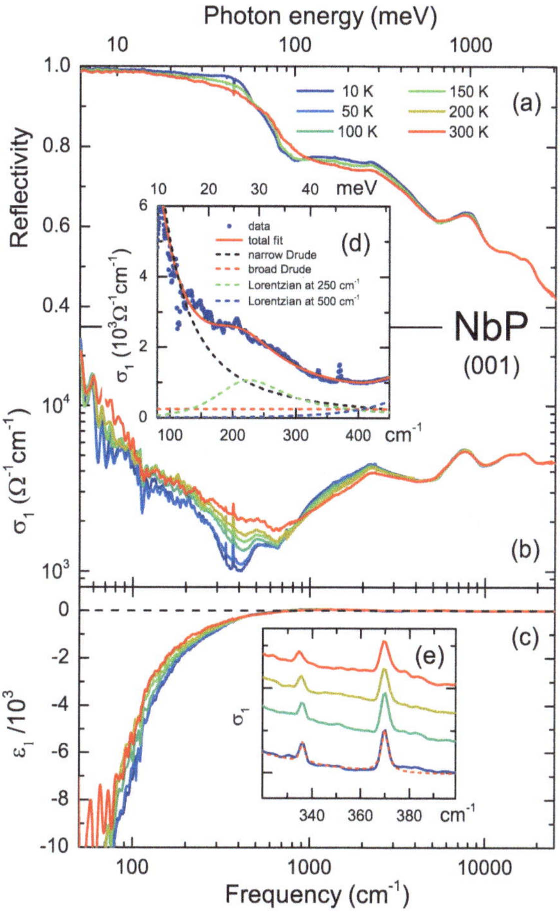

Fig. 3.13 Optical reflectivity (**a**), real parts of the optical conductivity (**b**) and dielectric permittivity (**c**) of NbP at selected temperatures between $T = 10$ and 300 K; note the logarithmic frequency scale. The inset (**d**) shows a simple fit of the low-energy $\sigma(\nu)$ at $T = 10$ K by the sum of two Drude terms (narrow and broad) and two Lorentzians centered at 250 and 500 cm^{-1}, which mimic interband transitions. The inset (**e**) displays the phonon modes in $\sigma(\nu)$ on an enlarged frequency scale. The dashed red line corresponds to a fit of the features at $T = 10$ K by two narrow Lorentzians centered at 336 and 370 cm^{-1}, respectively. Reproduced with permission from [174]. Copyright (2018) by the American Physical Society

3.3.3.1 Phonons

In the far-infrared range, two sharp phonon peaks can be seen in Fig. 3.13 at 336 and 370 cm^{-1}. Due to their symmetric shape, they can be nicely fitted with Lorentzians, as demonstrated in panel (e). This is in contrast to the highly asymmetric phonon resonances observed in the optical measurements of TaAs, where, unlike in our measurements on NbP, the probing radiation propagated along the low-symmetry [107] and [112] crystallographic directions [232].

In NbP, four infrared-active phonons are expected [233]; however, there is no full consensus on the calculated frequencies [233, 234]. The phonon positions observed in our spectra agree very well with the calculations from [234] as well as with the Raman data presented there (in non-centrosymmetric structures, same phonon modes can be both, infrared and Raman, active). Thus, following [234], we assign the observed features at 336 and 370 cm^{-1} to those lattice vibrations that mainly involve the light P atoms. The other two infrared-active phonon modes are apparently too weak to be resolved on the electronic background.

3.3.3.2 Itinerant Charge Carriers

At the lowest frequencies, NbP exhibits an optical response typical for metals, i.e. the itinerant carriers dominate: the reflectivity approaches unity, $\varepsilon_1(\omega)$ is negative and diverges as $\omega \to 0$, $\sigma_1(\omega)$ exhibits a narrow zero-frequency peak, as seen in Fig. 3.13a–c. Panel (d) clearly shows a shoulder on this peak that can be fitted by a Lorentzian; it will be discussed in the next section.

The increase in $\sigma_1(\omega)$ at $\omega \to 0$ is large [note logarithmic vertical scale in panel (b)], but still the values of $\sigma_1(\omega)$ at our lowest frequency are below the dc values, cf. the inset of Fig. 3.12. This means that the optical conductivity experiences a drastic increase at the frequencies below our measurement window. Such behavior of $\sigma_1(\omega)$ is expected in materials with high carrier mobility and has been reported, e.g., in TaAs [235], $YbMnBi_2$ [231], and ZrSiS, see the previous chapter. We can hence conclude that the ultrahigh mobility of NbP [17, 21] provides a narrow Drude-like mode in $\sigma_1(\omega)$. This mode can indeed be fitted with a simple Drude model using the measured dc-conductivity values as zero-frequency extrapolations.

To make the analysis more accurate, we used a Drude-Lorentz approach [75] to fit $\hat{\sigma}(\omega)$ at low energies. We found that two Drude terms and two Lorentzians can accurately describe the spectra at frequencies $\leq$500 cm^{-1}, Fig. 3.13d [at higher frequencies, Lorentzians are not an adequate description of the interband transitions, as will be discussed further].

In NbP, four bands cross the Fermi level, see [216, 217] and also our calculations below. Nevertheless, two Drude terms are sufficient to account for the contribution of itinerant carriers into optical conductivity. Utilizing more Drude terms will only lead to ambiguities in determination of the fit parameters. Such minimalist approach is common in optical-conductivity studies and has widely been used for different multi-band systems [204, 210, 220, 231]. For NbP, the two-Drude approach is justified by the effectively two-channel conduction in this compound [17]. Highly mobile carriers in crossing linear bands manifest themselves as a narrow Drude mode (small scattering rates), while low-mobile carriers in parabolic bands contribute as a broad Drude term (larger scattering rates). Such two-channel optical conductivity seems to be natural for (topological) semimetals and it also relevant, e.g., for YbPtBi, as discussed in the next chapter (however, it does not work for Au_2Pb, as shown in the previous section).

From the fits, we can determine the (momentum-relaxing) scattering rate γ_{mr} of the narrow Drude term, i.e. of the highly mobile carriers. The broad Drude term vastly overlaps with the interband transitions, thus determination of the scattering rate of low-mobile carriers is impossible. We have found γ_{mr} to be as low as 4.5 cm^{-1} at $T = 10$ K. The corresponding scattering time is $\tau_{mr} = 1/(2\pi\gamma_{mr}) = 1.2$ ps and the momentum-relaxation length, $\ell_{mr} = v_F\tau_{mr}$, becomes as long as 0.2 to 0.6 μm. Here we utilized the lower, 1.5×10^5 m/s, and, respectively, the upper, 4.8×10^5 m/s, boundaries for the (001)-plane Fermi velocity v_F obtained from experiment [17, 21] and theory [216, 217]. At elevated temperatures, γ_{mr} rises, reaching 35 cm^{-1} at $T = 300$ K. This corresponds to ℓ_{mr} of 20 to 70 nm. Let us note that because the scattering rates of the narrow Drude term are below our measurement frequency window, the vales obtained in this paragraph can only be considered as order-of-magnitude estimates.

3.3.3.3 Interband Transitions

Calculations of the interband optical conductivity based on calculated electronic band structure are very useful, but seem to be rather challenging in (topological) semimetals. A survey of the available literature reveals only a qualitative match between the calculated optical conductivity and experimental results [220, 231, 236, 237]. In the most interesting low-energy part of the spectrum (less than a few hundred meV), a reasonable agreement is particularly hard to achieve [231, 236]. In the case of NbP, we have reached a fairly good qualitative match between our calculations of the optical conductivity and experimental spectra even at low energies; this allows us to identify the origin of various spectral features observed in the measured $\sigma_1(\omega)$.

In Fig. 3.14 we compare the experimental low-temperature optical conductivity to the interband (001)-plane conductivity calculated with and without SOC. Note, that the itinerant-carriers contributions have not been subtracted from the experimental conductivity. The calculated spectra qualitatively agree with the experiment: the inset in Fig. 3.14 illustrates that the peaks and dips in the calculated $\sigma_1(\omega)$ reasonably coincide with those in the experimental data. The calculated spectra contain more fine structures than the experimental $\sigma_1(\omega)$, as no broadening was applied to the computed spectra in order to simulate finite life-time effects.

Above an energy of 0.2 eV, the effect of SOC on the theoretical $\sigma_1(\omega)$ is negligible; but for smaller energies the spectra computed with and without taking SOC into account differ significantly. The interband contribution to the conductivity calculated without SOC increases smoothly when raising the photon energy to 0.1 eV.

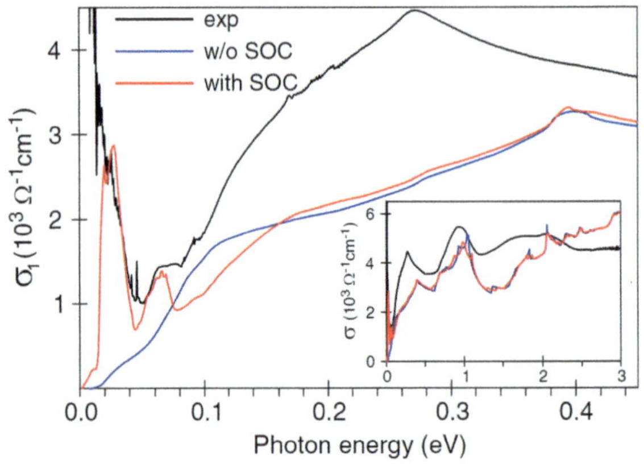

Fig. 3.14 Real part of the interband optical conductivity of NbP calculated with (red line) and without (blue line) SOC and the total experimental NbP conductivity at 10 K (black line). Intraband (Drude) contributions to the conductivity are not included in the computations. Inset shows same sets of data on a broader photon energy scale (0–3 eV). All spectra are for the (001)-plane response. Reproduced with permission from [174]. Copyright (2018) by the American Physical Society

When SOC is included, however, two sharp peaks appear around 30 and 65 meV (corresponding to $\sim$250 and 500 cm^{-1}). The latter matches very well the shoulder we observed on the narrow Drude term, as shown in Fig. 3.13d. The former feature can be directly associated with the bump observed at all temperatures at around 500 cm^{-1} in the measured spectra plotted in Fig. 3.13b. The fact that these two peaks appear only in those calculations including SOC indicates that they must be related to the transitions between the SOC-split bands.

Our conclusion gets support when decomposing the calculated $\sigma_1(\omega)$ into contributions coming from transitions between different pairs of bands crossing E_F. When SOC is neglected, two doubly degenerate bands with predominant Nb d character cross on a $k_{x/y} = 0$ mirror plane and form one electron and one hole Fermi surface with crescent-shaped cross sections by this plane [216]. This degeneracy is lifted when SOC is accounted for; thus, four non-degenerate bands, numbered 19 to 22, now cross E_F as shown in Figs. 3.15 and 3.16 (at every given **k** point, the bands are numbered with increasing energy). The band structure of NbP, calculated along selected lines in the BZ [cf. Fig. 3.16b], and the allowed transitions between different bands are shown in Fig. 3.15a, b. Here, the thickness of the vertical lines connecting occupied initial and unoccupied final states is proportional to the probability of the interband transition at a given **k** point. In panel (c) of Fig. 3.15, we plot the various contributions to the total interband conductivity from all individual interband transitions: 19 $\to$ 20, 19 $\to$ 21, 19 $\to$ 22, 20 $\to$ 21, 20 $\to$ 22, and 21 $\to$ 22.

The bands can be characterized by their spin polarization $\langle s \rangle$ which is well defined, i.e. $\langle s \rangle \simeq \pm 1/2$, only for **k**-vectors faraway from the band crossing points. Closer to the band crossings, SOC effects are strong and spin polarization is much less perfect. Thus, transitions between any pair of bands are allowed here. The transitions 19 $\to$ 21

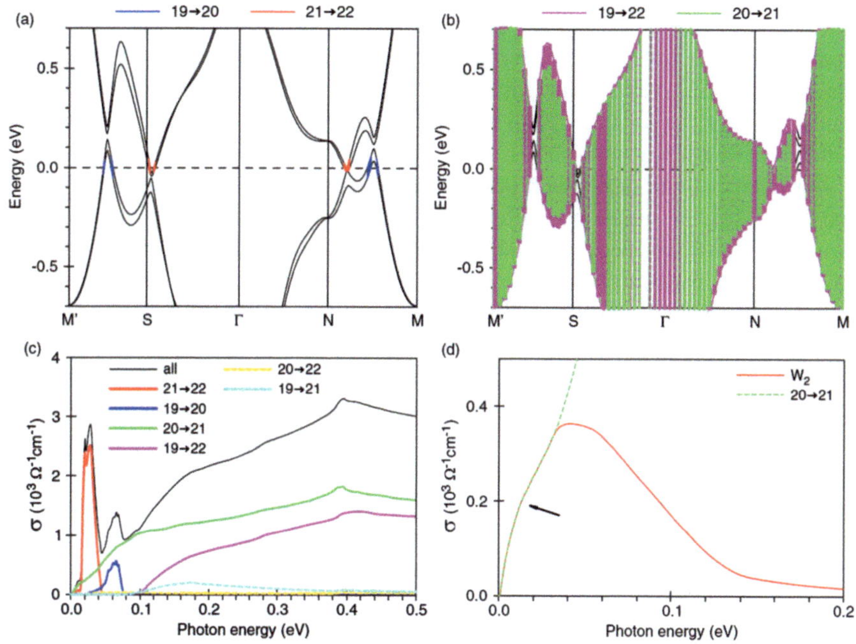

Fig. 3.15 Panels **a** and **b**: Band structure of NbP along selected lines in the BZ with allowed transitions shown as colored vertical lines. The thickness of the lines is proportional to the transition probability. Panel **c**: Calculated contributions to the interband conductivity from transitions between different pairs of bands crossing E_F, as indicated, and the total calculated interband conductivity of NbP. Panel **d**: Contribution to $\sigma_{20\to21}(\omega)$ from the transitions within the small volumes in **k**-space, enclosing the W2 Weyl points (red solid line, see text for details), in comparison with the total $\sigma_{20\to21}(\omega)$ (dashed green line). The arrow indicates a kink, which corresponds to the point, where the chiral Weyl bands merge. Reproduced with permission from [174]. Copyright (2018) by the American Physical Society

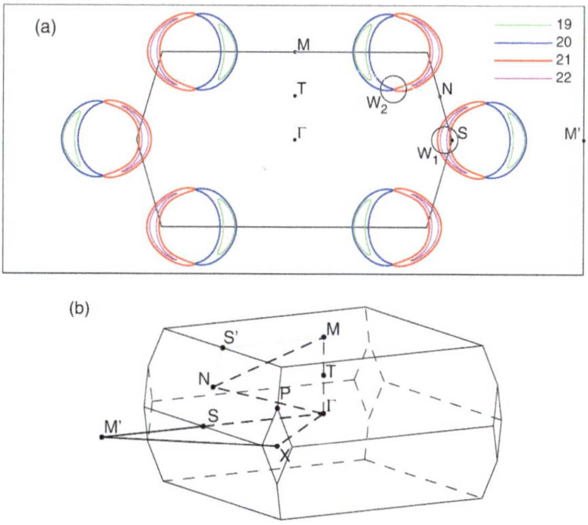

Fig. 3.16 **a** Fermi surface cross sections by $k_y = 0$ plane. Small black circles illustrate integration volumes around Weyl points. **b** BZ of NbP. Weyl points from the W1 set are situated near the S point, while the points from the W2 set are close to the N–M line. Reproduced with permission from [174]. Copyright (2018) by the American Physical Society

and 20 → 22 (the bands in each pair here possess opposite spin characters away from the nodes) still provide very small contributions to the optical conductivity [dashed cyan and yellow lines in Fig. 3.15c] because the allowed low-frequency transitions near the nodes are Pauli-blocked and the transition probability diminishes with increasing frequency.

For other pairs of bands with predominantly opposite spin characters (19 → 20 and 21 → 22), the situation is different. The pronounced narrow peak at 30 meV is formed by transitions between the bands 21 and 22 [red line in Fig. 3.15c], while the 65 meV mode is due to the 19 → 20 transitions [blue line]; i.e. both features stem from transitions between SOC-split bands. These transitions are only allowed in a small volume of the **k**-space, where one of the two SOC-split bands is occupied, while the other one is empty (see the red and blue lines in Fig. 3.15a), i.e., between the nested crescents in Fig. 3.16a. Within this volume, the SOC-split bands are almost parallel to each other. This ensures the appearance of strong and narrow peaks in the joint density of states for the corresponding interband transitions. The peak positions are determined by the average band splitting, which, in turn, is of the order of the SOC strength of the Nb d states, $\xi_d \approx 85$ meV. Finally, since the dipole matrix elements for these transitions are rather large, the two peaks dominate the low-energy interband conductivity.

Besides these two peaks, there are also significant contributions to the optical conductivity with a smooth ω dependence. These contributions originate in transitions between the touching bands 20 and 21 [green lines in Fig. 3.15b, c] and between the bands 19 and 22, which are separated by a finite gap everywhere in the BZ [magenta lines]. Accordingly, $\sigma_{20\to21}(\omega)$ starts at zero energy, while $\sigma_{19\to22}(\omega)$ at 0.1 eV. Both conductivity contributions, $\sigma_{20\to21}$ and $\sigma_{19\to22}$, increase, when the photon energy rises from 0 to 0.4 eV. Both contributions, $\sigma_{19\to22}$ and $\sigma_{19\to22}$, as well as the total calculated $\sigma_1(\omega)$ exhibit sharp kinks (Van Hove singularities [79]) at 0.4 eV, which are related to the transitions between flat parallel bands near the N point, see Fig. 3.15b. The experimental $\sigma_1(\omega)$ demonstrates such a kink at somewhat lower energy, 0.27 eV (Fig. 3.14); still, we find this match reasonable.

In the vicinity of a Weyl point, the optical conductivity is expected to be proportional to frequency, (1.34). The two sets of Weyl points in NbP, W1 and W2, are formed by touching points of the bands 20 and 21. In agreement with previous results [216, 218–220], our LMTO calculations yield the W1 points approximately 50 meV below E_F. Consequently, their contribution to the conductivity cannot start at zero frequency [80]. On the other hand, the energy of the W2 points in the present calculations is very close to E_F and, thus, transitions near W2 may provide a linearly vanishing $\sigma_1(\omega)$ as $\omega \to 0$. To verify this behavior, we calculate the contribution to $\sigma_{20\to21}(\omega)$ from a **k** volume with a radius of $\sim 0.05\frac{2\pi}{a}$ (a is the in-plane lattice constant) around the averaged position of a pair of W2 points. The contribution indeed shows a linear ω dependence as $\omega \to 0$, see Fig. 3.15d. The smooth kink at ~15 meV, marked with an arrow, corresponds to the merging point of the chiral Weyl bands [79, 83]. We should note that in experiments, this linear interband optical conductivity at low ω is completely masked by the itinerant carriers and by the strong peaks due to the transitions between the SOC-split bands, as discussed above.

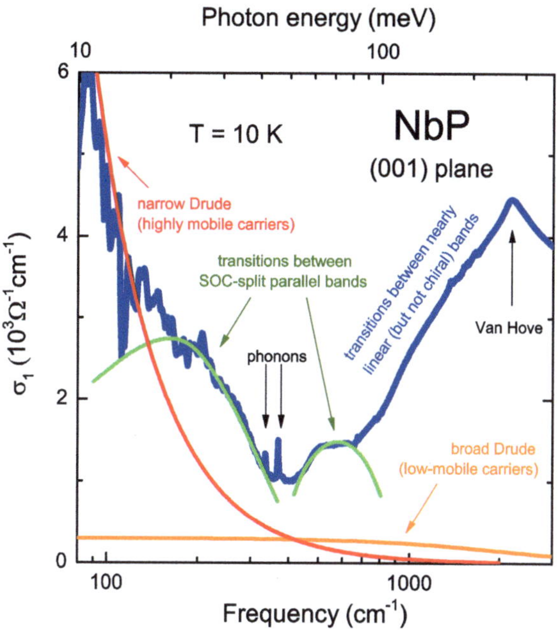

Fig. 3.17 Low-frequency portion of the real part of NbP optical conductivity at 10 K and assignment of the observed features to different absorption mechanisms, as discussed in this chapter. Note that the broad-Drude and Lorentzian terms are shown schematically; the scattering rate of the broad Drude mode is ill defined. Reproduced with permission from [174]. Copyright (2018) by the American Physical Society

Although no linear-in-frequency $\sigma_1(\omega)$ due to the transitions within the chiral Weyl bands can be seen in NbP at low frequencies, both, experimental and computed, $\sigma_1(\omega)$ demonstrate a sort of linear increase with ω at higher frequencies: 180–250 meV for the experimental and up to 360 meV for the calculated optical conductivity, see Fig. 3.14. As apparent from our calculations, the linearity just reflects the fact that all the electronic bands, involved in the transitions with corresponding energies, are roughly linear (but not parallel to each other), see Fig. 3.15b.

Based on the comparison between the calculated and the experimental conductivity, we can assign the observed spectral features to different absorption mechanisms. Figure 3.17 schematically summarizes these assignments.

Before we conclude, we would like to emphasize the importance of the transitions between the SOC-split bands. So far, the strong influence of these transitions on the low-energy conductivity of Weyl semimetals has not been fully appreciated. Using a modified Dirac Hamiltonian [5], Tabert and Carbotte [83] calculated the optical conductivity for a four-band model, relevant for many Weyl semimetals. In this model, the band structure consists of four isotropic non-degenerate 3D bands, two of which cross and the other two are gapped. The band structure is mirror symmetric in energy with respect to the Weyl nodes and the Weyl cones are anisotropic in **k**-space at low energies. This model definitely grasps the main features of the band structure of many Weyl semimetals, including those from the TaAs family; but—apart from neglecting possible band anisotropy—it does not take into account the transitions between the SOC-split bands: these transitions are considered forbidden in the model, while in the real Weyl semimetals they might play an important role, as we have shown for NbP.

Finally, we would like to note that TaAs and TaP demonstrate sharp absorption peaks at frequencies, which compare well to those of the transitions between the SOC-split bands in NbP, cf. Fig. 3 from [236]. Previously, these features have been assigned to transitions between the merging (saddle) points of the Weyl bands, even though, e.g., in TaP this assignment is at odds with the absence of chiral carriers [108]. We suggest to reconsider this assignment.

3.3.4 Conclusions

We measured and analyzed the interband and free-carrier optical conductivity of NbP. From the electronic band structure, we calculated the interband optical conductivity and decomposed it into contributions from the transitions between different bands. By comparing these contributions to the spectral features in the experimental conductivity, we assigned the observed features to certain interband transitions. We argued that the low-energy (below 100 meV) interband conductivity is dominated by transitions between almost parallel bands, split by spin-orbit coupling. Hence, these transitions manifest themselves as relatively sharp peaks centered at 30 and 65 meV (240 and 525 cm^{-1}). These peaks and the low-energy free-carrier conductivity (Drude-like) conceal the linear-in-frequency contribution to $\sigma_1(\omega)$ from the transitions within the chiral Weyl bands. Our calculations demonstrate that the nearly linear in ω conductivity at around 200–300 meV (1600–2400 cm^{-1}) is naturally explained by the fact that all electronic bands, involved in the transitions with such frequencies, possess approximately linear dispersion relations. We also identify two optical phonons and assign them to the vibrations, which mostly involve P atoms. Finally, we find that the itinerant carriers at least in one of the conduction channels possess very low momentum-relaxing scattering rates at low temperatures, leading to mesoscopic characteristic lengths (hundreds of nanometers) of momentum relaxation.

3.4 The Weyl Semimetal TaP

This section is based on our earlier work on TaP [176]. Similarly to NbP, TaP is a nonmagnetic non-centrosymmetric Weyl semimetal from the TaAs family with 24 Weyl nodes of two different types, commonly dubbed as W1 (8 nodes) and W2 (16 nodes) [16, 18, 215, 216, 220, 238]. The available band-structure calculations predict that the W1 nodes are situated some 40–55 meV below the Fermi level E_F, while the W2 nodes are at 12–20 meV above it [216, 220, 238].

As was reported in the previous sections of this chapter, in many real Dirac or Weyl semimetals, the anticipated linear interband conductivity at low energies is (partly) masked by other features, such as intraband (Drude) conductivity or

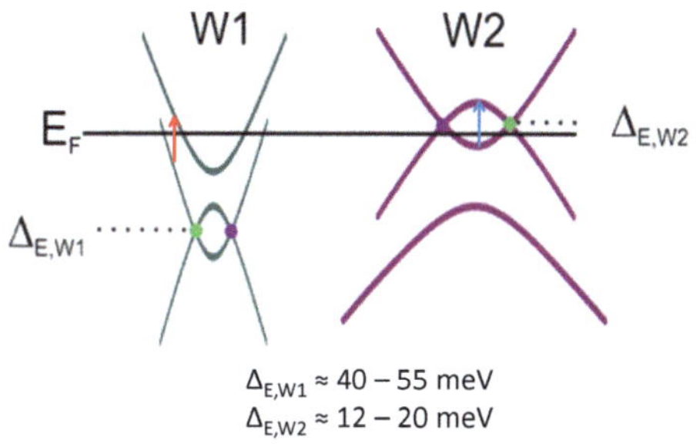

Fig. 3.18 A schematic diagram of the Weyl bands in TaP. Possible transitions between the saddle points of the merging Weyl bands and between the SOC-split bands are indicated as arrows. Reprinted from [176] under CC-BY 4.0 licence. Copyright (2021) by the authors

resonance-like interband contributions. Particularly strong low-energy peak was observed in TaP [175, 236]. In [236], this peak was assigned to electron-hole pair excitations near the saddle points of the crossing bands, which form the Weyl nodes (Fig. 3.18). To our knowledge, this assignment doesn't have a direct support from optical-conductivity calculations based on band structure. Also, the total number of states near the saddle points is relatively low. Hence, only relatively small kinks in the optical conductivity, rather than strong peaks, are expected in this situation [83, 174]. On the other hand, our study of the sister compound NbP has demonstrated that the low-energy peaks, similar to the one in TaP, appear in NbP and are due to multiple transitions between almost parallel bands split by SOC. Based on our band structure calculations, we argue here that the same explanation of its low-energy peak holds also for TaP.

3.4.1 Results and Discussion

In Fig. 3.19 we plot our experimental optical spectra. The measurements have been done on the isotropic *ab* plane of TaP (cf. Fig. 1.4) in the same way as the measurements of NbP. The prominent low-energy peak is clearly seen in the real part of the optical conductivity at 50 meV (it corresponds to a dip in the optical reflectivity).

To gain insight into the origin of this peak, we performed band structure calculations within the local density approximation (LDA) based on the experimental crystal structure of TaP [239]. Similarly to our calculations for NbP, we utilized the linear muffin-tin orbital (LMTO) method [228] with the Perdew-Wang exchange-correlation potential [229] and the relativistic PY LMTO computer code [227] with SOC added to the LMTO Hamiltonian in the variational step. BZ integrations were done using the improved tetrahedron method [230]. Additional empty spheres (E) were inserted at the 8b Wyckoff positions in order to minimize the effect of atomic sphere overlap. The Ta, P, and E states up to the maximal orbital quantum number $l_{max} = 3, 2, 1$, respectively, were included into the LMTO basis set, which is essential for calculations of the dipole matrix elements for the interband transitions involving the Ta d- and the P p-derived bands. When calculating the real part of the

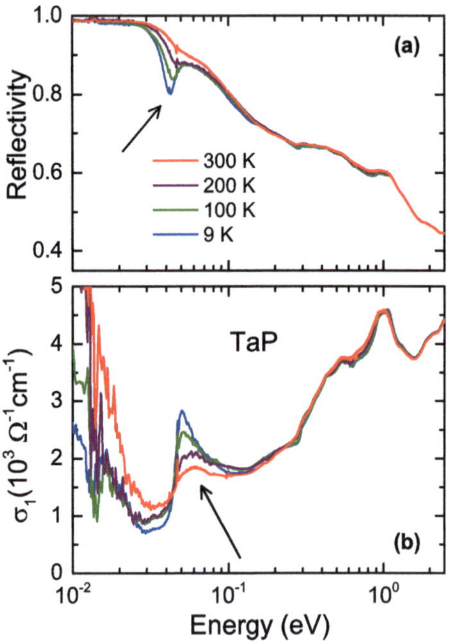

Fig. 3.19 Experimental in-plane reflectivity (**a**) and the corresponding real part of the optical conductivity (**b**) of TaP at selected temperatures. The arrows mark the feature discussed here. The increased σ_1 at low energies is due to intraband (Drude-like) absorption. Reprinted from [176] under CC-BY 4.0 licence. Copyright (2021) by the authors

optical conductivity, we used the tetrahedron method on a dense $128 \times 128 \times 128$ k-mesh in order to resolve interband transitions between the SOC-split bands close to Weyl points. No broadening has been applied to the computed spectra.

In Fig. 3.20 we show the results of our band-structure calculations as well as the BZ of TaP. Four non-spin-degenerate bands, numbered 19 to 22, can be resolved in the vicinity of E_F (at every given $\mathbf{k}$ point, the bands are numbered with increasing energy staring from the lowest calculated band). Note that the bands in each of the two pairs, (19, 20) and (21, 22), are split by SOC because of the lack of inversion symmetry. Our results well reproduce the available literature data [216, 220].

Before we discuss the calculated optical conductivity spectra, we would like mention that the unintentional (self-)doping (due to the crystallographic defects, impurities, and vacancies) may slightly change the position of E_F. Such unintentional doping, varying from sample to sample, has been shown to be relevant to TaP [240]. On the other hand, band structure calculations themselves have finite accuracy (cf. the spread in the calculated energy positions of the Weyl nodes, mentioned above). These considerations justify small variation of the position of E_F to get a better match between theory and experiment. Hereafter, we vary the Fermi-level position within $\Delta E_F = \pm 100$ meV.

Figure 3.21 presents the results of our interband-conductivity calculations starting from the self-consistent band structure shown in Fig. 3.20. The black solid line in Fig. 3.21 shows $\sigma_1(\omega)$ obtained with the as-calculated E_F. The overall run of the experimental interband conductivity is well reproduced by this curve: $\sigma_1(\omega)$ increases

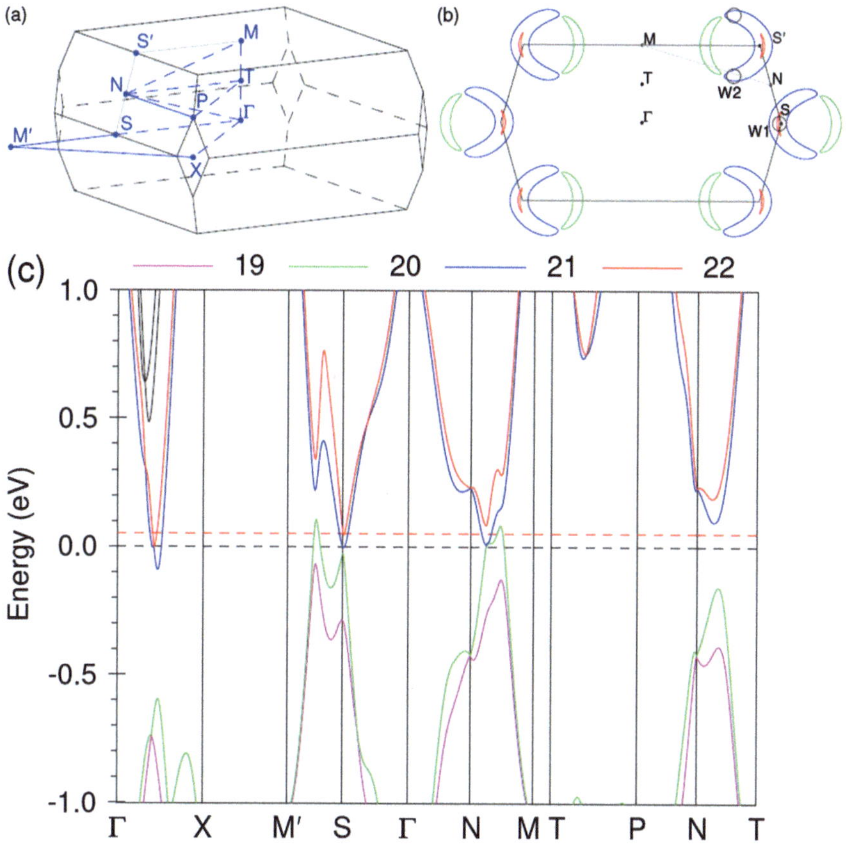

Fig. 3.20 **a** The BZ of TaP. **b** Fermi surface cross sections calculated for $\Delta E_F = 50$ meV. Black circles show approximate positions of projections of Weyl points onto $k_y = 0$ plane. **c** Band structure of TaP. Four bands closest to E_F (marked 19–22) are shown in different colors. Black and red horizontal dashed lines show the as-calculated position of E_F and the Fermi level shifted upwards by $\Delta E_F = 50$ meV, respectively. Reprinted from [176] under CC-BY 4.0 licence. Copyright (2021) by the authors

with frequency and reaches a maximum at 1 eV (cf. Fig. 3.19 and note that the intraband (Drude) contribution has not been taken into account in the band-structure computations). Nevertheless, no peak is visible in these computations at around 50 meV. A slight variation of E_F provides such a peak, but only if the Fermi level is shifted upwards (red and blue curves). Shifting E_F downwards does not change the σ_1 spectra in the desirable way (magenta and cyan curves). The hight of the 50-meV peak reaches the experimental value of $2.5 \times 10^3\ \Omega^{-1}\text{cm}^{-1}$ at $\Delta E_F = 50$ meV.

In the inset in Fig. 3.21, we present an expanded view of the low-frequency optical conductivity calculated for small positive ΔE_F. It is obvious, that even a very small shift of E_F by 20 meV is sufficient to produce the 50-meV peak. Note, that for all

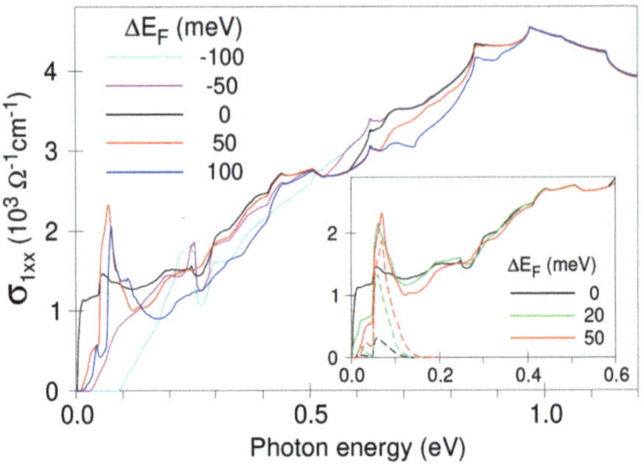

Fig. 3.21 Low-energy optical conductivity of TaP calculated from its band structure. Lines of different colors correspond to different positions of the Fermi level, as indicated. The conductivity calculated for smaller positive ΔE_F is plotted in the inset. The contributions of $21 \rightarrow 22$ transitions are shown by dashed lines. Reprinted from [176] under CC-BY 4.0 licence. Copyright (2021) by the authors

three curves yet another experimental feature—a broad shoulder at 0.3–0.5 eV—is also evident in the calculated spectra. Thus, we can conclude that a tiny shift of the Fermi level allows one to obtain a very reasonable overall description of the experimental $\sigma_1(\omega)$, including the strong peak at 50 meV.

To understand what interband optical transitions are responsible for this peak, one can take a look at Fig. 3.20, where the original and shifted by 50 meV Fermi levels are shown by the dashed black and red lines, respectively. In the vicinity of Weyl points, i.e., near the S point and along the N–M line, band 21 is above the as-calculated E_F. The low-frequency interband conductivity is dominated by the transitions between the initial band 20 and the final band 21. The shift of E_F to higher energy leads to partial occupation of band 21. This suppresses the $20 \rightarrow 21$ transitions at low energy and, at the same time, allows transitions from band 21 to band 22, which remains mostly empty. As these SOC-split bands are almost parallel, the energies of such transitions are expected to be roughly the same for different momenta. Thus, a sharp peak may occur in $\sigma_1(\omega)$.

To confirm this observation, we performed band-resolved optical-conductivity calculations for the transitions between bands 21 and 22. The results of these calculations for three ΔE_F are plotted by dashed lines in the inset of Fig. 3.21. It is apparent, that the $21 \rightarrow 22$ transitions provide the major contribution to the 50-meV peak, confirming our proposition. A $21 \rightarrow 22$ contribution coming from the **k** volume near the middle of the Γ–X line also appears for the as-calculated E_F, but it is too weak to be responsible for the 50-meV peak.

Here, we would like to emphasize the importance of the transitions between the SOC-split bands. Such transitions can be considered forbidden in some models [83], while in the real Weyl semimetals they play an important role, as we have shown above for NbP. These transitions are allowed, because the electronic bands can be characterized by their well-defined spin polarization, $\langle s \rangle \simeq \pm 1/2$, only for $\mathbf{k}$-vectors faraway from the Weyl nodes; closer to the nodes, SOC is strong and spin polarization is much less perfect. Thus, transitions between any pair of bands are allowed there.

3.4.2 Conclusions

Summarizing, we have calculated low-energy optical conductivity of the Weyl semimetal TaP (in the ab plane) and compared it to the experimental results. The best match between theory and experiment is found for a slightly shifted Fermi level (+20–60 meV). This shift confirms a small unintentional doping of TaP, discussed earlier [175, 240], and offers a natural explanation of the strong low-energy (50 meV) peak reported in the experimental data [175, 236]: the peak is due to transitions between almost parallel non-degenerate electronic bands split by spin-orbit coupling. This is very similar to the situation in NbP and can be relevant to other Weyl semimetals from this family.

3.5 Chiral Anomaly in Weyl Semimetals

3.5.1 Chiral Anomaly in TaAs

3.5.1.1 Chiral Anomaly in Steady Fields

The experiential detection of the chiral-anomaly-induced features in optics is challenging, because these features are not expected to be huge. In the experiments described in this section, we tried to detect the chiral anomaly in steady electric and magnetic fields by using thin flakes of the Weyl semimetal TaAs in transmission experiments. We used a high-power free-electron laser (FLARE) and a Bitter magnet, available at Radboud University in Nijmegen, the Netherlands [241]. The measurements described here were performed at 21 cm^{-1} (2.6 meV) [242]. The light was partly polarized with an approximately 60:40 percent intensity ratio. The sample's transmission was measured in a magnetic field of 20 T in a tilted Faraday geometry, as shown in Fig. 3.22 (the tilt was necessary to provide $\mathbf{B} \cdot \mathcal{E} \neq 0$). Fairly high ($\pm$100 mA) dc currents had to be utilized for this semimetallic material to create a reasonable electric field inside it. To prevent possible heating and suppress impurity scattering, the sample was submerged in helium and its temperature was kept at 1.6 K. The sample was a (001)-plane oriented single crystal of TaAs with lateral dimensions of roughly 2.5 by 2.5 mm and a thickness of 200 micron. The crystal was synthesized via chemical vapor transport, as described in [108, 243].

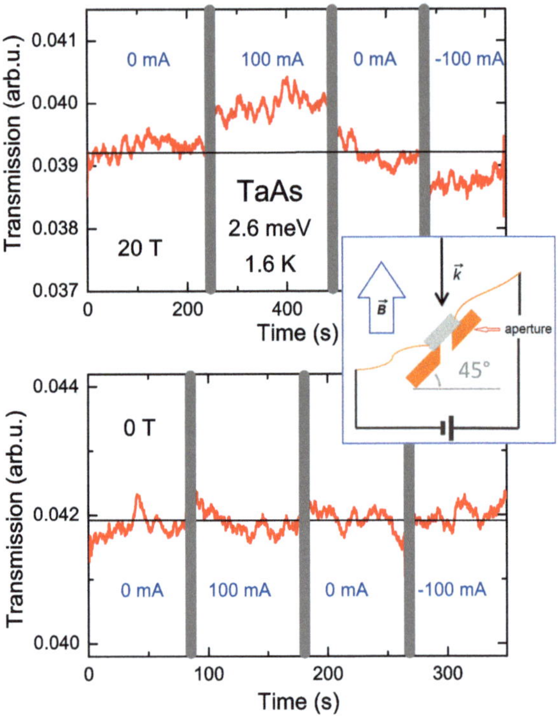

Fig. 3.22 An attempt to detect the chiral anomaly in TeAs in steady electric and magnetic fields: Transmission through the TaAs sample in 20 T (upper frame) and in 0 T (lower frame) as a function of time. The applied currents are indicated. The inset schematically shows the measurement geometry. Reprinted from [242] under CC-BY 4.0 license. Copyright (2019) by the authors

We have chosen to perform transmission measurements on thin single crystals using a very powerful radiation source, rather than more standard reflectivity measurements, because TaAs is highly reflecting in the THz and far-infrared frequency regions relevant for the chiral Weyl bands [111, 235, 236]. Hence, possible chiral-anomaly-induced changes would be extremely hard to detect on this high-reflectivity background.

The major results of this attempt are shown in the main panels of Fig. 3.22. It is evident from the figure, that there is an effect on the optical transmission upon simultaneous application of $\mathcal{E}$ and **B**: the effect is only present if $\mathbf{B} \neq 0$ and, most interestingly, it reverses its sign if $\mathcal{E}$ changes the direction. This can be related to the high elipticity of the probing light: the circular-polarized light absorption is supposed to change depending on the sign of $\mathcal{E} \cdot \mathbf{B}$ [113].

We should however note that at 20 T TaAs is basically in the quantum limit [238]. Thus, the low-field approximation used in [80, 113] is not directly applicable. It is still to be confirmed, whether the observed effect is (entirely) due to the chiral anomaly. For example, measurements with circular polarized light, rather than with only partial light polarization, as in the present experiment, are desirable. Nevertheless, it is evident that our observations are in line with the expectations for the optical signatures of the chiral anomaly.

3.5.1.2 Dynamic Chiral Anomaly

As discussed in Sect. 1.3.2, a more promising way to detect the chiral anomaly is to observe the changes in the chemical potential, induced by simultaneous application of static **B** and dynamic **e** fields. For TaAs, we conducted such measurements in the fields of up to 7 T and for both, ordinary and extraordinary, Voigt geometries. The TaAs sample of approximately 2 by 2 mm in lateral dimensions was kept at 10 K, and optical reflectivity between 30 and 650 cm^{-1} was measured from the (001) surface.

The results of these measurements are presented in Fig. 3.23. Here, the relative reflectivity, $R(B)/R(0)$, is plotted versus frequency. It is evident from the left panel of the figure, that in the ordinary Voigt geometry, which is supposed to be featureless for conventional materials, we do observe an appearance of a bump developing at around 120 cm^{-1}, cf. Fig. 1.14. The hight of the bump growths, as B rises. This is perfectly in line with the expectations for the dynamic chiral anomaly. We note, that

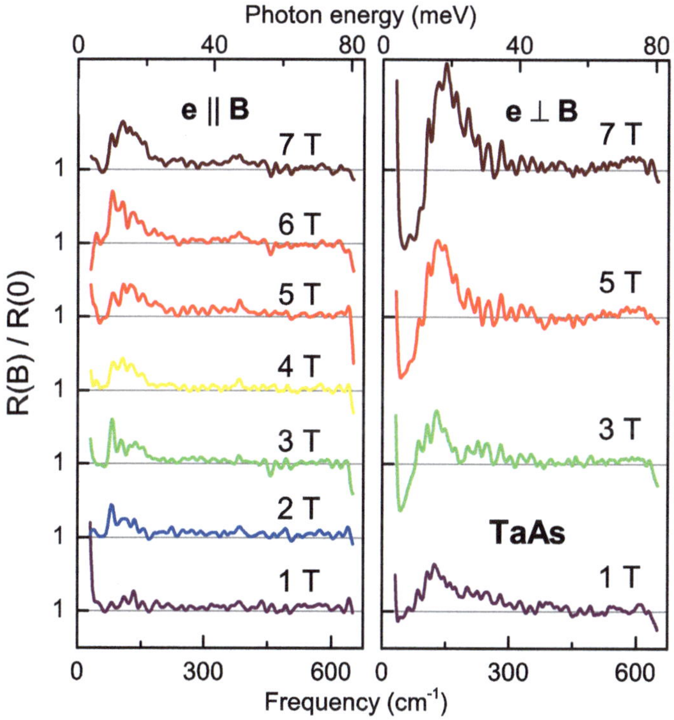

Fig. 3.23 Left panel: results of the magneto-optical measurements of TaAS in the ordinary Voigt geometry. The appearance of the bump indicates the dynamic chiral anomaly. Right panel: results in the extraordinary Voigt geometry. The cyclotron resonance splits the plasma edge, confirming its frequency position. The dashed horizontal lines mark the unity of the $R(B)/R(0)$ of each field indicated. The curves shifted upwards with increasing B field

the plasma edge of our TaAs crystal is situated right at these frequencies. Thus, the field-induced changes do reflect the dynamic chiral pumping.

This conclusion is also confirmed by our measurements in another Voigt geometry (right panel of Fig. 3.23). Here, **e** $\perp$ **B** and a cyclotron resonance is expected to split the plasma edge. This very behaviour we observe: the $R(B)/R(0)$ ratio first drops and then rises, while the frequency increases and passes through the (zero-field) plasma frequency.

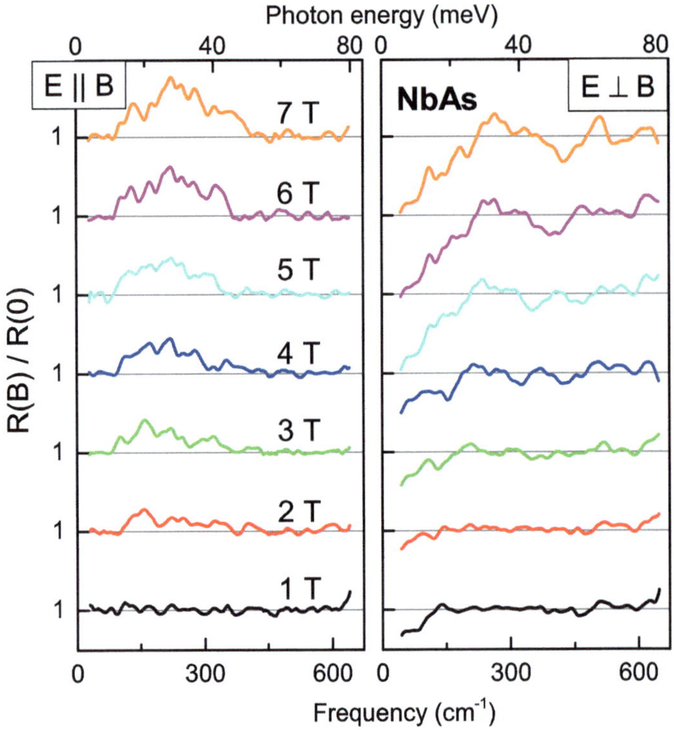

Fig. 3.24 Chiral anomaly in NbAs. Left panel: results of the magneto-optical measurements of NbAs in the ordinary Voigt geometry. Note that E on the plots stands for the electric-field component of the probing light. The appearance of the bump and its shift to higher frequencies with increasing magnetic field indicate the dynamic chiral anomaly. Right panel: results in the extraordinary Voigt geometry. The cyclotron resonance splits the plasma edge, confirming its frequency position. The dashed horizontal lines mark the unity of the $R(B)/R(0)$ of each field indicated. The curves shifted upwards with increasing B field

3.5.2 Dynamic Chiral Anomaly in NbAs

Very similar results on the dynamic chiral anomaly have been obtained by us in another Weyl semimetal, NbAs. A graphical summary of our results is given in Fig. 3.24. The measurements were performed on the (001) surface at 10 K. Let us note that the extraordinary-Voigt spectra are slightly distorted by inter-Landau-level transitions. The measurements for $\mathbf{e} \parallel \mathbf{B}$ are not affected by these transitions and demonstrate a nice peak, which is less noisy than the related peak in the TaAs spectra. It can be noticed that the peak shifts to higher frequencies, as magnetic field increases, which is indeed expected for the dynamic chiral anomaly.

To conclude this section, we can claim optical observation of the (dynamic) chiral anomaly in two Weyl semimetals—TaAs and NbAs. Our results on TaAs confirm the previous observation [111], while the data for NbAs are new. Overall, the presence of the chiral bands is confirmed by optical means in these materials.

Chapter 4
Triple-Point Semimetals

4.1 GdPtBi—Broadband Optical Response

4.1.1 Introduction

Heusler materials are currently well recognized for their wide range of spectacular electronic and magnetic properties [244]. The high tunability of these compounds allows designing materials with properties on demand for future functioning applications [245, 246]. Recently, band inversion and topologically non-trivial electronic states have been intensively studied in (half)Heusler compounds with strong spin-orbit coupling [247–254].

Among other half-Heuslers with strong SOC, GdPtBi occupies a special place: hallmarks of a Weyl-semimetal state, such as negative magnetoresistance and planar Hall effect, are vividly developed in this material and are assigned to manifestations of the chiral anomaly [106, 109, 255]. It has been proposed that the band structure of GdPtBi in zero magnetic field can be sketched as two degenerate parabolic bands touching each other at the Γ point of the BZ [255]. A moderate external magnetic field splits the bands. This leads to linear-band crossings and a WSM state, which enables negative longitudinal magnetoresistance [109, 255] and planar Hall effect [106]. Most recent density-functional-theory calculations, though, forecast linear-band crossings even in zero magnetic field [256]. These nodes are, however, different from the Dirac and Weyl points: in the GdPtBi case, one doubly and one non-degenerate bands cross each other, forming the so-called triple points [25, 257]. Angular-resolved-photoemission reveals linear electronic bands in GdPtBi, but these bands are mostly assigned to the surface states [251]. As mentioned earlier, optical methods are more sensitive to bulk properties because of large penetration depths. Therefore, it is tempting to probe the optical response of GdPtBi and to compare it with theory predictions.

Here, we present measurements of the optical conductivity in GdPtBi (this study was published in [258]). We found $\sigma_1(\omega)$ to be linear in a broad frequency range: at

A. V. Pronin, *Linear Electrodynamic Response of Topological Semimetals*,
Springer Series in Solid-State Sciences 199,
https://doi.org/10.1007/978-3-031-35637-7_4

$T \leq 50$ K, the linearity spans from ~100 meV down to a few meV. We calculated $\sigma_1(\omega)$ from the GdPtBi band structure and, by comparing the experimental and the computed conductivity, demonstrated that the linear-in-frequency $\sigma_1(\omega)$ is due to electronic transitions between the bands in the vicinity of the triple points.

4.1.2 Sample Preparation and Experimental Details

GdPtBi single crystals were grown by the solution method from a Bi flux. Freshly polished pieces of Gd, Pt, and Bi, each of purity larger than 99.99%, in the ratio Gd:Pt:Bi = 1:1:9 were placed in a tantalum crucible and sealed in a dry quartz ampoule under 3 mbar partial pressure of argon. The filled ampoule was heated at a rate of 100 K/h up to 1200 °C, followed by 12 h of soaking at this temperature. For crystal growth, the temperature was slowly reduced by 2 K/h to 600 °C. Extra Bi flux was removed by decanting it from the ampoule at 600 °C. Overall, the crystal-growth procedure followed closely the ones described in [254, 259]. Crystals' composition and structure (non-centrosymmetric $F\bar{4}3m$ space group) were checked by energy dispersive X-ray analysis and Laue diffraction, respectively.

Our optical reflectivity measurements were conducted on a single crystal with lateral dimensions of ~2×1.1 mm^2 and with a shiny (111) surface (inset in Fig. 4.1); the sample thickness was around 0.8 mm. Standard four-point dc-resistivity and Hall measurements, performed on a smaller piece (a Hall bar) cut from the specimen used for optics, indicated a semiconducting behavior with a well-known antiferromagnetic transition at 9 K [259]. Hall measurements show the p-type conduction and a very low carrier density of 6×10^{17} cm^{-3} at $T \rightarrow 0$ (cf. different samples from [255]).

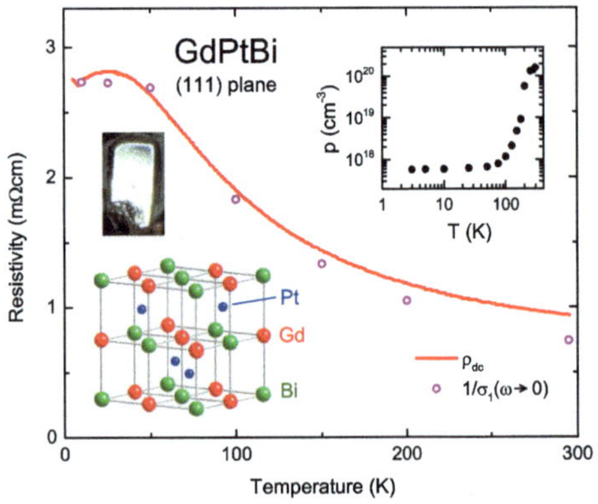

Fig. 4.1 Temperature-dependent dc resistivity (solid line) of GdPtBi and the inverse values of its optical conductivity at $\omega \rightarrow 0$ (open dots). Inset: carrier concentration p obtained from Hall measurements. The sample used in this work is also shown alongside with the GdPtBi structure. Reproduced with permission from [258]. Copyright (2018) by the American Physical Society

The results of our transport measurements are presented in Fig. 4.1. All optical experiments reported here have been made in the paramagnetic state ($T \geq 10$ K), where the Dirac physics of GdPtBi is primarily discussed [106, 109, 255, 256].

Optical reflectivity $R(\nu)$ was measured at 10–300 K over a frequency range from $\nu = \omega/(2\pi c) = $ 20–22000 cm^{-1} (2.5 meV–2.73 eV) using two Fourier-transform infrared spectrometers, as described in Chap. 3. Complex optical conductivity was obtained from $R(\nu)$ using Kramers-Kronig transformations. High-frequency extrapolations were made utilizing the x-ray atomic scattering functions [145]. At low frequencies, we used the same procedure as for ZrSiS: we fitted the $R(\nu)$ spectra with a set of Lorentzians and then used the results of these fits between $\nu = 0$ and 20 cm^{-1} as zero-frequency extrapolations for subsequent Kramers-Kronig transformations.

4.1.3 Experimental Results

Figure 4.2 displays an overview of the results obtained in our optical investigations. Panel (a) shows the reflectivity for all measurement temperatures. Panels (b) and (c)

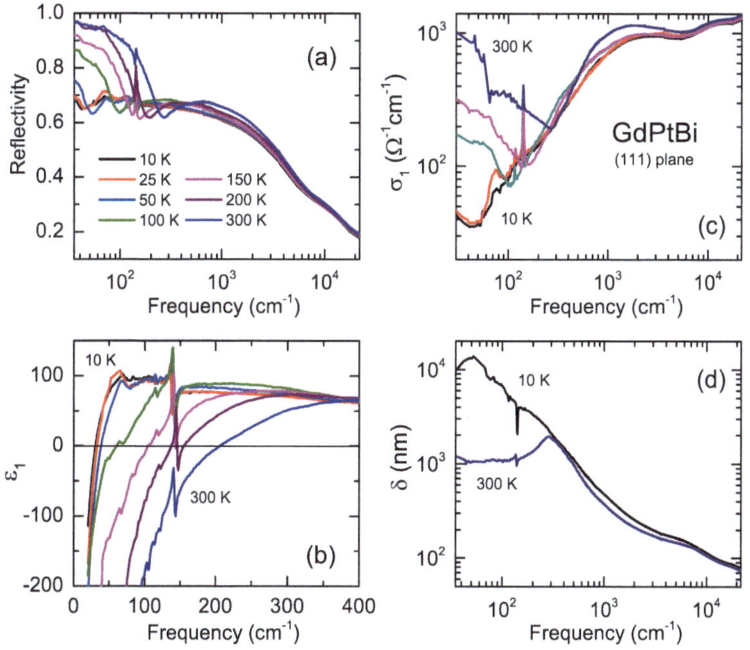

Fig. 4.2 Reflectivity (**a**), dielectric constant ε_1 (**b**), optical conductivity σ_1 (**c**), and skin depth δ (**d**) of GdPtBi at different temperatures as indicated. The low-frequency portion of $\sigma_1(\nu)$ is magnified in Fig. 4.3. Reproduced with permission from [258]. Copyright (2018) by the American Physical Society

represent (the real parts of) the dielectric constant, $\varepsilon_1(\nu)$, and the optical conductivity, $\sigma_1(\nu)$, respectively. Finally, panel (d) demonstrates the skin depth $\delta(\nu)$ of our sample at 10 and 300 K (curves for intermediate temperatures lie in-between of these ones). Important is that the skin depth exceeds 50 nm for all measured temperatures and frequencies. In the most interesting low-energy region, it is in the micrometer range. Hence, our optical measurements reflect bulk properties.

A sharp phonon peak is seen in all data sets at $\sim$140 cm^{-1}. Another phonon at $\sim$115 cm^{-1} is week, but resolvable, especially in $\varepsilon_1(\nu)$ [panel (b)]. The frequency positions of both phonon modes have only marginal temperature dependence. No other phonons are detected, in agreement with group analysis, which predicts two infrared-active optical modes for the half-Heusler structure [260–263]. All other features of the optical response are due to intra- or inter-band electronic transitions, as discussed right below.

A temperature-dependent plasma edge dominates the low-energy part ($\nu < 300$ cm^{-1}) of the reflectivity spectra [panel (a)]. The edge corresponds to the screened plasma frequency of free carriers, $\nu_{\rm pl}^{\rm scr}$, [75] and is also seen in panel (b) as zero crossings of the $\varepsilon_1(\nu)$ curves. From the same panel, it can also be noted that the background dielectric constant ε_∞ is rather high, around 70–100. This leads to a low unscreened plasma frequency, $\nu_{\rm pl} = \nu_{\rm pl}^{\rm scr}\sqrt{\varepsilon_\infty}$ (for example, $\nu_{\rm pl} \approx 300\,{\rm cm}^{-1}$ at 10 K), in agreement with the low free-carrier concentration found in Hall measurements. The free-electron contribution to the optical conductivity [panel (c)] is seen as a Drude-like mode at the lowest frequencies. At lower temperatures, this mode narrows and looses its spectral weight in accordance with decreasing $\nu_{\rm pl}^{\rm scr}$ at $T \to 0$. As $T \to 0$ K, only marginal traces of the free-carrier (intra-band) contribution are seen in the recorded $\sigma_1(\nu)$ spectra: above $\sim$50 cm^{-1}, $\sigma_1(\nu)$ reflects only the interband optical transitions (and the phonons, as mentioned above).

A striking feature of the optical conductivity is its almost perfect linearity in a broad range in the far-infrared. This can be seen best in Fig. 4.3, where experimental conductivity is shown alongside with linear fits [square-root behavior of conductivity, expected for parabolic bands, is also shown for comparison]. The behavior of experimental $\sigma_1(\nu)$ is basically the same for the three lowest measurement temperatures (10, 25, and 50 K): it linearly increases with ν in the spectral range from approximately 50 up to almost 800 cm^{-1}. Observation of this linearity is an important result of this work. As discussed above, the linearity of the low-energy $\sigma_1(\nu)$ is a signature of 3D linear bands [90, 91, 170, 172, 235, 236]. However, other band structures may also provide similar $\sigma_1(\nu)$. For example, it can be a cumulative effect of many bands with predominantly, but not exclusively, linear dispersion relations. Such situation was discussed earlier in this book for the Weyl semimetal NbP. Wavy deviations from a perfectly linear increase of $\sigma_1(\nu)$ [see, Fig. 4.3b and c] indicate that a similar scenario might be realized in GdPtBi.

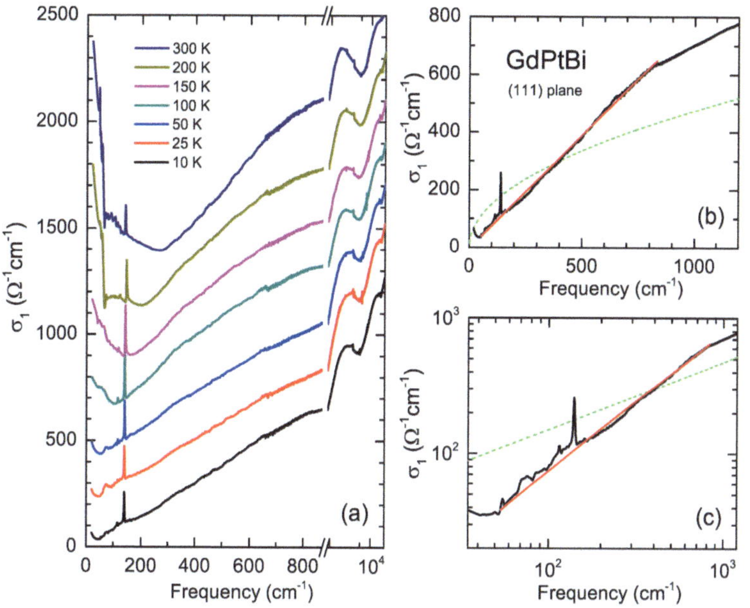

Fig. 4.3 Optical conductivity of GdPtBi with emphasis on low frequencies (**a**). Note a frequency-scale change at 900 cm^{-1}. The curves for $T \geq 25$ K are shifted upwards for clarity (by 200 Ω^{-1}cm^{-1} as compered to the previous measurement T). Linear fit (straight red line) of the experimental $\sigma_1(\nu)$ at 10 K for $50 < \nu < 800$ cm^{-1} on linear (**b**) and log-log (**c**) scales. A square-root behavior of $\sigma_1(\nu)$, expected for parabolic bands, is shown by dashed green lines. Reproduced with permission from [258]. Copyright (2018) by the American Physical Society

4.1.4 Computations and Analysis

To get an insight into the origin of the linear $\sigma_1(\nu)$, we performed band-structure calculations and then computed the interband optical conductivity. In the calculations, we used the linear muffin-tin orbital method (LMTO) [228] as implemented in the relativistic PY LMTO computer code [227, 264]. The Perdew-Burke-Ernzerhof GGA [265] was used for the exchange-correlation potential. The $4f^7$ states of gadolinium were treated as semi-core states. The $4f$ spin-polarization was not considered in order to model the paramagnetic state studied in this work ($T \geq 10$ K). Relativistic effects, including SOC, were accounted for by solving the 4-component Dirac equation inside atomic spheres. BZ integrations were done using the improved tetrahedron method [230].

The calculated band structure is shown in Fig. 4.4. Our calculations confirm the presence of basically parabolic bands, touching each other at the Γ point. A closer look on the low-energy band structure, Fig. 4.5a, b, reveals the presence of a triple point (marked as 3p in the figure) along the $\Gamma - L$ line, in agreement with previous calculations [256]. The bands in the plane normal to the $\Gamma - L$ direction possess linear dispersion relations, as shown for the 3p $-$ X direction in panel (b). There

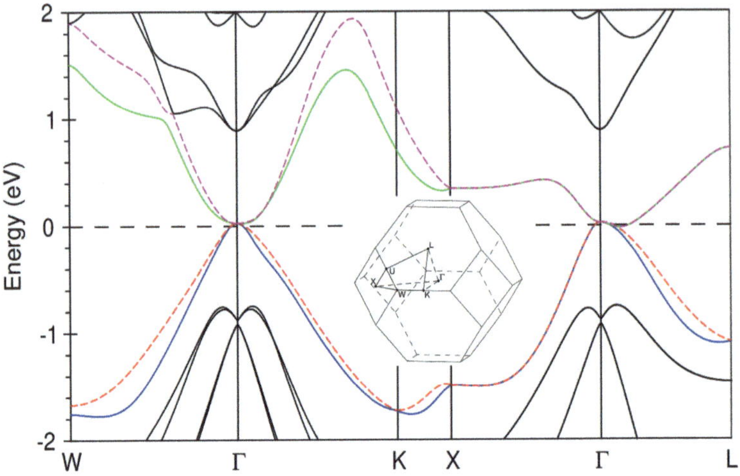

Fig. 4.4 Band structure of GdPtBi. Low-energy bands are displayed in different colors. BZ is shown as inset. Reproduced with permission from [258]. Copyright (2018) by the American Physical Society

are 8 symmetry-equivalent triple points per BZ. The band structure of GdPtBi near these points is similar to a Weyl or Dirac semimetal with tilted cones, as seen best in panel (a).

Our calculations predict the triple points to be situated 18 meV below the Fermi level. However, the real GdPtBi crystals are known to often possess unintentional doping, which is impossible to control on the crystal-growing stage [255]. Thus, the position of the chemical potential can in reality be within a few tens of meVs off the calculated value. It is instructive to note here that the linear-in-frequency $\sigma_1(\nu)$ is expected for tilted cones (of any type), if the chemical potential μ is situated at the node [81]. In practice, 2μ (μ is measured form the node hereafter) should be below the measurement frequency window. Such situation can be relevant for our GdPtBi sample, as we discuss below. This is also in agreement with the very small free-carrier (Drude) contribution and low Hall carrier density.

The band structure of GdPtBi in the vicinity of Fermi level is obviously more complex than the model band structure used in [81]. Thus, as mentioned above, we compute $\sigma_1(\nu)$ from the obtained band structure. In these computations, the dipole matrix elements for interband optical transitions were calculated on a $96 \times 96 \times 96$ k-mesh using LMTO wave functions—it is necessary to use sufficiently dense meshes in order to resolve transitions at low energies [231]. The real part of the optical conductivity was calculated using the Kubo–Greenwood linear-response expressions [266] with the BZ integration performed using the tetrahedron method.

The results are shown in Fig. 4.5c and d. Because of the possible carrier doping in GdPtBi, discussed above, we have some freedom in setting the position of the chemical potential. We varied μ within ± 30 meV from the tripe point and compared

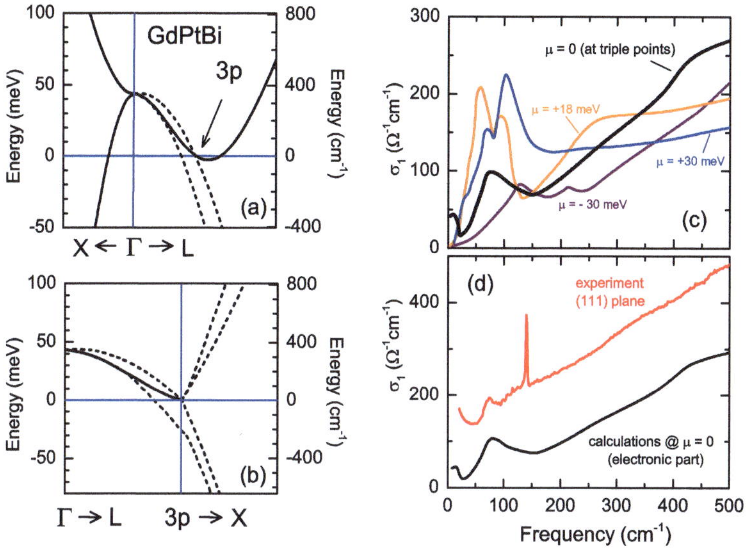

Fig. 4.5 Panels **a** and **b**: Electronic bands of GdPtBi near the triple point (marked as 3p). The chemical potential μ is set to zero at the triple point. Doubly degenerate bands are shown as solid lines, while non-degenerate bands as dashed lines. Panel **c**: Calculated interband conductivity of GdPtBi, $\sigma_1(\omega)$, for a few different positions of μ as indicated. Panel **d**: Comparison of the measured (upper curve) and calculated for $\mu = 0$ (bottom curve) optical conductivity of GdPtBi. Phonon modes are not included in the calculated $\sigma_1(\omega)$. The experimental curve is shifted upwards by $100\,\Omega^{-1}\mathrm{cm}^{-1}$ for clarity. Reproduced with permission from [258]. Copyright (2018) by the American Physical Society

the computed $\sigma_1(\nu)$ spectra to each other and to the experiment. Panel (c) demonstrates that the best linear $\sigma_1(\nu)$, extrapolating to 0 at $\nu \to 0$, is obtained, if the chemical potential is at the triple point ($\mu = 0$). If we vary μ, the calculated $\sigma_1(\nu)$ either develops huge peaks at low energies ($\nu < 200\ \mathrm{cm}^{-1}$), or does not extrapolate to 0 as $\nu \to 0$, or both. Also, the quasi-linear part of the conductivity, calculated for $\mu = 0$, spans over the largest frequency range. Thus, we choose the $\mu = 0$ curve for further comparison with our experimental results, see panel (d). (Obviously, very small deviations from $\mu = 0$ on a meV scale are possible.).

In Fig. 4.5d, a low-temperature (25 K) experimental curve is shown alongside with the calculated $\sigma_1(\nu, \mu = 0)$. The overall linear increase of the experimental curve is well reproduced. It is also evident that both, calculations and experiment, provide some deviations from perfect linearity. Most remarkable is the bump, present in the calculations and experiment, at around $80\ \mathrm{cm}^{-1}$. Such deviations reflect the fact that the band structure is not ideally linear in all three directions, but more complex. Overall, we can conclude that the observed interband optical conductivity in GdPtBi originates from the transitions between all the bands near the triple points. Linear terms dominate the dispersions of these bands in the close vicinity of the nodes, leading to the almost, but not perfectly, linear optical conductivity in GdPtBi at low frequencies.

From our band-structure calculations, we can compute the Fermi velocities v_F for the crossing bands. Calculations exactly at the triple point are technically challenging, thus we compute v_F in a close vicinity of it along the $\Gamma - L$ line – at $\pm 0.005 \times 2\pi/a$ from the triple point; here a is the lattice constant. For the doubly degenerate electron-like band, we obtain $v_F = 1.1$ and 0.5×10^5 m/s, while for the non-degenerate hole-like band $v_F = 2.4$ and 2.8×10^5 m/s.

In the simple model of electron-hole symmetric crossing linear bands, the optical conductivity is related to the Fermi velocity v_F via (1.33). Let us slightly modify it to be able to describe the triple-point semimetals:

$$\sigma_1(\omega) = \frac{e^2 g N}{24h} \frac{\omega}{v_F}. \tag{4.1}$$

Here, g is the band degeneracy at the crossing point (e.g., a Dirac node has $g = 4$) and N is the number of nodes per BZ. Obviously, this simple formula has a very limited applicability. Nevertheless, if we straightforwardly apply it to our experimental $\sigma_1(\omega)$ and set $g = 3$ and $N = 8$, we obtain an averaged Fermi velocity of $\sim 10^5$ m/s, which is in a good agreement with the values calculated above.

4.1.5 Conclusions

Summarizing, we have found the low-frequency optical conductivity of GdPtBi to be linear in a broad frequency range (50–800 cm^{-1}, ~6–100 meV at $T \leq 50$ K). This linearity strongly suggests the presence of three-dimensional linear electronic bands with band crossings near the chemical potential. Comparison of our data with the optical conductivity computed from the band structure demonstrates that the observed $\sigma_1(\omega)$ originates from the transitions near the triple points. From the optical spectra, we directly determine the plasma frequency of free carriers in GdPtBi and estimate an averaged Fermi velocity at the nodes, $v_F \sim 10^5$ m/s. The values of v_F, calculated from the band structure near the triple points along the $\Gamma - L$ line, range from 0.5 to 2.8×10^5 m/s depending on the band and the momentum direction.

4.2 Chiral Anomaly in GdPtBi

A magnetic-field-induced Weyl semimetal state has been suggested in GdPtBi by Hirschberger et al. [255]. This state originates from the Zeeman splitting of the spin-degenerate band forming the triple point. This field-induced state has been backed by careful measurements of negative magnetoresistance [109].

Here, we show that GdPtBi, similarly to TaAs and NbAs (Sects. 3.5.1 and 3.5.2), clearly demonstrates the optical signatures of the dynamic chiral anomaly. We performed measurements on the same sample and for the same orientation as described

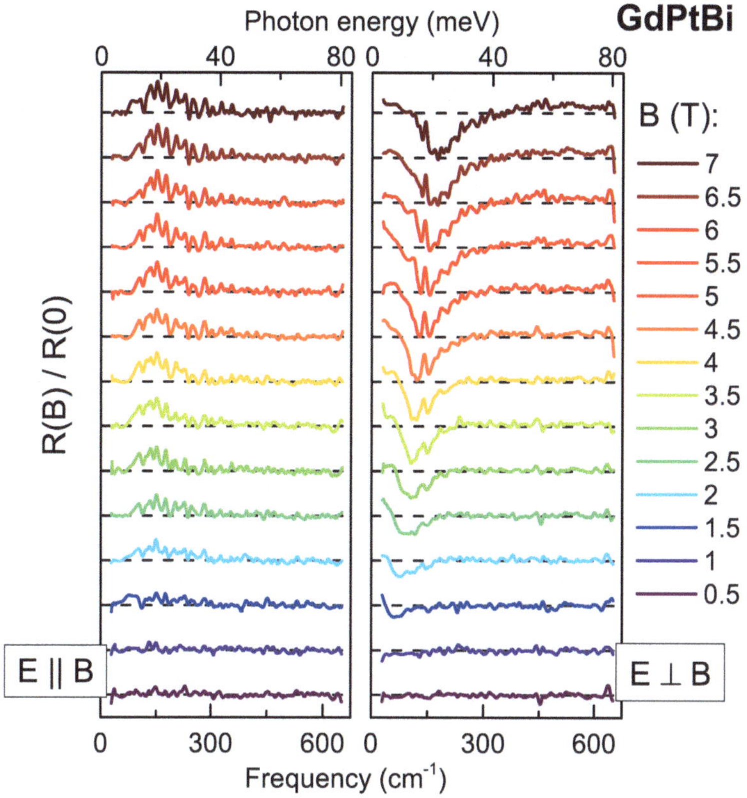

Fig. 4.6 Chiral anomaly in GdPtBi. Left panel: results of the magneto-optical measurements of GdPtBi in the ordinary Voigt geometry. The appearance of the bump and its growth with increasing magnetic field indicate the dynamic chiral anomaly. Right panel: results in the extraordinary Voigt geometry. The dashed horizontal lines mark the unity of the $R(B)/R(0)$ of each field indicated. The curves shifted upwards with increasing B field. Note that E on the plots stands for the electric-field component of the probing light (marked as **e** in the main text)

above in this Chapter. The measurements have been made in the Voigt geometry with the external magnetic field **B** and the electric field of the probing electro-magnetic wave $\mathbf{e}(t)$ parallel to each other. For comparison, measurements for another Voigt orientation, $\mathbf{B} \perp \mathbf{e}(t)$, have also been performed. The B field was changed from 0 to 7 T. The sample temperature was about 20 K, i.e., GdPtBi was in the paramagnetic state. We measured the ratio of the reflectivity in applied field $R(B)$ to the zero-field reflectivity $R(0)$ in the range from 30 to 650 cm^{-1}.

In Fig. 4.6, a graphical summary of our measurements is presented. Here, the relative reflectivity, $R(B)/R(0)$, is plotted versus frequency. In the left panel, the results for the $\mathbf{B} \parallel \mathbf{e}$ case are shown. It is evident, that in the ordinary Voigt geometry, there is small, but detectable, bump developing at around 140 cm^{-1}, cf. Fig 1.14. The hight of the bump growths, as B rises. This is perfectly in line with the expectations for the dynamic chiral anomaly. In the extraordinary Voigt geometry (right panel of Fig. 3.23), we observe a dip in the $R(B)/R(0)$ spectra. This dip is likely related to a CR mode, but it doesn't show up in a typical CR way, likely because of a very high anisotropy of the Fermi pockets in GdPtBi. Still, we can conclude that our observations of the dynamic chiral anomaly in GdPtBi confirm the magnetic-field induced Weyl-semimetal state in this compound.

4.3 YbPtBi

4.3.1 Introduction

For 25 years YbPtBi was renowned as a heavy-fermion compound that exhibits one of the highest effective electron masses among the strongly correlated electron systems [259, 267]. Its Kondo temperature is around 1 K; in addition an antiferromagnetic transition is observed at 0.4 K. So far, most of the experimental studies on YbPtBi explore its heavy-fermion state and a possible quantum critical point; hence they focus on temperatures below 2 K [259, 267–273].

More recently, however, it has been emphasized that YbPtBi belongs to the large family of intermetallic ternary, so-called half-Heusler, compounds, which demonstrate a variety of rather interesting electronic properties. Several half-Heuslers are predicted to exhibit band inversion at the Γ point, leading to topologically non-trivial states [247, 248, 250, 254]. Similarly to GdPtBi, YbPtBi may also possess triple points in its band structure. Let us note that due to the combination of diverse electronic properties and non-trivial band topology, half-Heuslers are currently recognized as extremely promising objects in the research towards functioning materials. This calls for a more comprehensive look on the electronic properties of YbPtBi, beyond the heavy-fermion state.

Here, we concentrate on the temperature range well above the Kondo temperature, i.e. 2.5–300 K, where from our magnetotransport and optical measurements we can draw conclusions on the carrier dynamics. We present evidence that two types of charge carriers coexist in this compound: highly mobile electrons with a temperature-dependent carrier concentration of the order of 10^{18} cm^{-3} and holes with a rather high and basically temperature-independent concentration of 10^{20} cm^{-3} that possess a very low mobility. We find the values of electron mobility in YbPtBi to be record high for half-Heuslers. The presence of the highly mobile carriers is typical for materials with linear bands [14, 17]. Possible presence of such bands in YbPtBi has been noticed e.g. in [247], but still remains an open issue. The found very high mobility in YbPtBi indicates that Dirac physics might indeed be relevant for this compound, however more studies in this regard are certainly necessary. Our results are published in [146].

4.3.2 Experiment

YbPtBi single crystals were grown by the solution growth method, where Bi acts as a flux. Stoichiometric quantities of freshly polished pieces of elements Yb, Pt, and Bi of purity >99.99% in the atomic ratio of 0.7:0.7:10 were put in a tantalum crucible and then sealed in a dry quartz ampoule under 3–5 mbar argon pressure. The filled ampoules were heated at a rate of 100 K/h up to 1473 K, followed by 12 h of soaking; after that the furnace temperature was decreased to 1373 K. For crystal growth, the temperature was slowly reduced from 1373 K to 873 K by 2 K/h and the surplus of Bi flux was removed by decanting the ampoule at 873 K. Using this method, we obtained 3–5 mm regular triangular shaped crystals, with a preferred growth in the (111)-direction. The general crystal growth procedure was followed from literature [274]. The space group and lattice parameters are found to be $F\overline{4}3m$ (cubic face-centered) and 6.591 Å, respectively, consistent with previous reports [275, 276].

Specimens of appropriate shapes were cut from a single crystal, e.g. Hall bars for (magneto)transport experiments and a large-surface sample for optical reflectivity. Direct-current resistivity measurements were performed in a custom-made setup by cooling from room temperature down to 2.5 K. Transversal magnetoresistance (MR) and Hall resistivity measurements were conducted at the same temperatures in magnetic fields B of up to 6.5 T. The voltages/currents were measured/applied within the (111) plane and magnetic field was along the [111] axis.

The optical reflectivity $R(\nu)$ was measured from the (111) plane in the temperature range between $T = 12$ and 300 K using two Fourier-transform infrared spectrometers, a Bruker IFS 113v and a Bruker Vertex 80v equipped with an infrared microscope. This way we covered the frequency range from 100 to 22 000 cm^{-1} ($\hbar\omega = 12$–2700 meV). The sample for the optical experiments had lateral dimensions of roughly 2 by 2 mm and a typical thickness of 0.5 mm; all optical experiments were performed on freshly cleaved (111) surfaces. As usual, for low frequencies (50–1000 cm^{-1}), an in-situ gold evaporation technique [192] was utilized for reference measurements and freshly evaporated gold and protected-silver mirrors served as references at higher frequencies. In accord with the cubic face-centered crystal structure, measurements with linearly polarized light revealed isotropic optical properties.

From the measured reflectivity $R(\nu)$, the optical conductivity and dielectric function, were extracted via the Kramers-Kronig relations. Here, we express our optical results in terms of $\sigma_1(\nu)$ and $\varepsilon_1(\nu)$. For a Kramers-Kronig analysis, the measured data have to be extrapolated to zero and high frequencies. Similarly to the situation with ZrSiS and NbP, it turned out that in the case of our measurements the commonly applied Hagen-Rubens extrapolation to zero frequency was not adequate because a very narrow Drude component is present in the spectra with a scattering rate comparable to our lowest measurement frequency, $\nu_{\text{min}} \approx 100\,\text{cm}^{-1}$. Thus, we used a set of Lorentzians instead—some at a finite, some at zero frequency—to fit the measured reflectivity. Between $\nu = 0$ and $\nu_{\min}$ these Drude-Lorentz fits were utilized as zero-frequency extrapolations. On the high-frequency side ($\nu \to \infty$), the x-ray atomic scattering functions were employed according to Tanner [145].

4.3.3 Results and Discussion

4.3.3.1 Magnetotransport

The dc resistivity ρ_{dc} and conductivity, $\sigma_{dc} = 1/\rho_{dc}$, are displayed in Fig. 4.7 versus temperature. Upon cooling, $\rho_{dc}(T)$ continuously decreases; the residual resistivity ratio is 8, comparable to the values reported earlier [267, 270]. In $\rho_{dc}(T)$, a point of inflection, tentatively attributed to the influence of crystalline-electric-field effects [270], is observed around $T = (90 \pm 5)$ K.

Figure 4.8 shows the results of our magnetotransport experiments by plotting the field-dependent resistivity taken at different temperatures. The MR curves are supplemented by data of Mun et al. [270] taken at $T = 1$ K. At all temperatures the magnetoresistance sharply increases with B in the range of low magnetic fields. It then flattens out for elevated temperatures, eventually reaching almost 60% as $B \to 6.5$ T. Such sharp increase in MR at low fields and its high-field saturation can be interpreted as localization of carriers with low scattering rate (i.e. with high mobility) in the fields of ~0.5 T, while other type of carriers, with high scattering rates and low mobility, still provides the dc transport. For $T < 20$ K the magnetoresistance

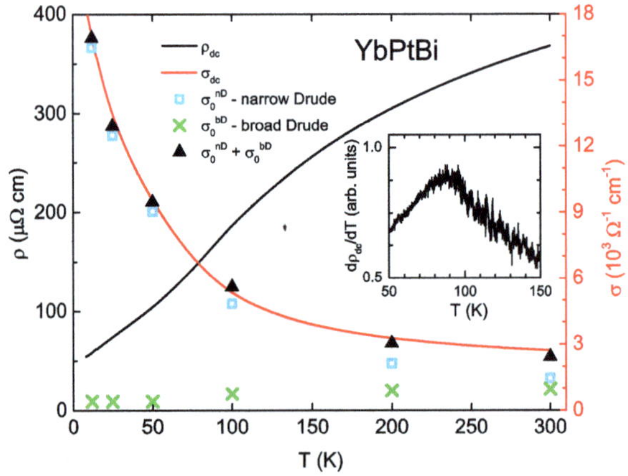

Fig. 4.7 Temperature-dependent four-point resistivity ρ_{dc} (black line, left scale) and dc conductivity, $\sigma_{dc} = 1/\rho_{dc}$, (red line, right axis) of YbPtBi. In addition, the bold dots correspond to the values obtained as $\nu \to 0$ extrapolations of the two Drude contributions, broad and narrow, to the optical conductivity, and to the sum of the two, see text. The inset shows $d\rho_{dc}/dT$. Reproduced with permission from [146]. Copyright (2017) by the American Physical Society

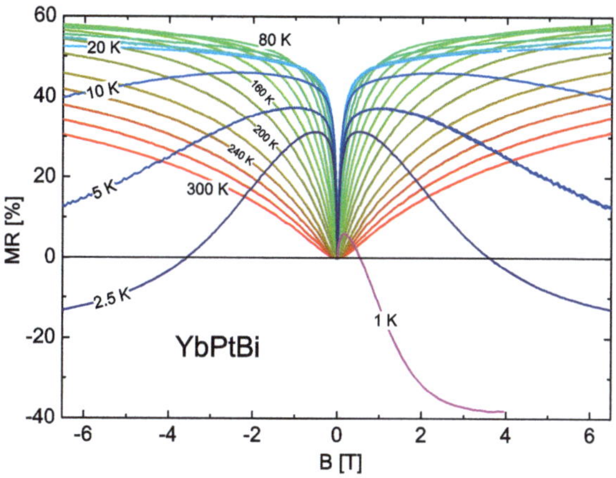

Fig. 4.8 Magnetoresistance of YbPtBi as measured between 2.5 and 300 K. The 1 K data are extracted from [270]. Reproduced with permission from [146]. Copyright (2017) by the American Physical Society

is not monotonic anymore: a pronounced maximum is observed that shifts to lower fields as temperature decreases. At very low temperatures, $T \leq 2.5$ K, MR changes sign: at $B = 3.5$ T for $T = 2.5$ K and around 0.5 T for the 1 K data of Mun et al. [270]. The position of the maximum in the MR-versus-B curves follows a power-law temperature dependence, T^n, with $n = 1.2 \pm 0.1$. Negative MR in YbPtBi is usually discussed in terms related to Kondo physics [267, 270], though no in-depth discission is available so far. It is worth to note here, that similar, but somewhat different, results on transversal MR are reported recently for related compounds, ScPtBi [277] and HoPdBi [278]. In ScPtBi, no negative MR is observed in the whole range of measured fields (0−10 T) and temperatures (2−300 K), but the shape of the MR-versus-B curves is very similar to our higher-temperature results. In HoPdBi, the negative MR sets in already at 50 K in 7 T and the overall shape of the lower temperature MR curves are similar to our results for $T < 20$ K, but the temperature dependencies of the MR maximum is different.

The results of the Hall measurements on YbPtBi, i.e. the Hall resistance, $R_{xy}(B, T)$, and the Hall coefficient, $R_H(B, T) = \rho_{xy}/B$, are plotted in Fig. 4.9 as the functions of applied magnetic field for various temperatures as indicated. The low-field behavior is magnified in the inset in order to demonstrate that R_H is always negative in low fields. As the field increases, the Hall coefficient eventually changes sign: electrons get localized and the Hall response becomes dominated by holes (accordingly, the MR saturates, as discussed). Within our range ($B < 6.5$ T), R_H turns positive for all temperatures below 260 K and saturates at 0.015 cm^3C^{-1}. Obviously, this very value would be reached at all temperatures up to 300 K, if we could go to higher magnetic fields, as demonstrated in the inset of the bottom panel. Thus, we conclude that the hole concentration n_h is temperature independent and equal to $(5.2 \pm 0.6) \times 10^{20}$ cm^{-3}.

Oppositely, the electron concentration n_e demonstrates a strong temperature dependence. We calculate $n_e(T)$ from $R_H = 1/(n_e \cdot e)$, taking into account only the initial slop in $\rho_{xy}(B)$ at low fields, and plot the results in Fig. 4.10a.

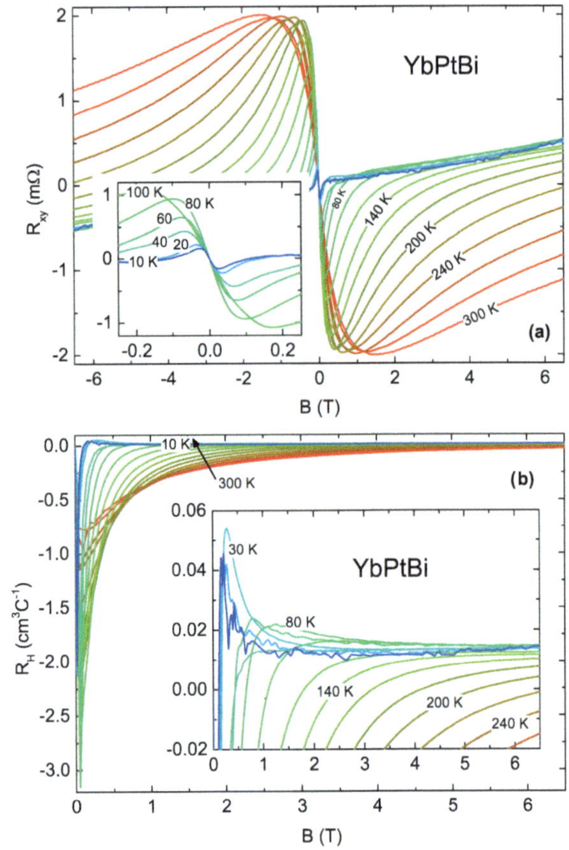

Fig. 4.9 Hall resistance (top panel) and Hall coefficient (bottom panel) of YbPtBi at 2.5–300 K. The insets zoom-in the Hall resistance at low fields (top panel) and the small values of the Hall coefficient (bottom panel). Reproduced with permission from [146]. Copyright (2017) by the American Physical Society

The two-carrier-type scenario for YbPtBi does not allow us to calculate the carrier mobility in a straightforward fashion because at this stage we cannot separate the contributions of each carrier type to the total conductivity σ_{dc}. In the so-called one-carrier-type (OCT) approach (i.e. if we use the total measured σ_{dc} to compute both, electron and hole, mobilities from the corresponding Hall coefficients), one can though estimate roughly the mobilities: $\mu_e^{\mathrm{OCT}} = R_H^e \cdot \sigma_{\mathrm{dc}} \approx 53\,000\ \mathrm{cm}^2/\mathrm{Vs}$ and $\mu_h^{\mathrm{OCT}} = R_H^h \cdot \sigma_{\mathrm{dc}} \approx 200\ \mathrm{cm}^2/\mathrm{Vs}$ at $T = 10$ K. This estimate not only provides the upper limits for the mobilities, but also shows that the mobilities of electrons and holes in YbPtBi differ from each other by orders of magnitude. Below we will describe, how our optical measurements provide additional information in order to separate the electron and hole contributions to the conductivity and thus to get more accurate values of the carrier mobilities.

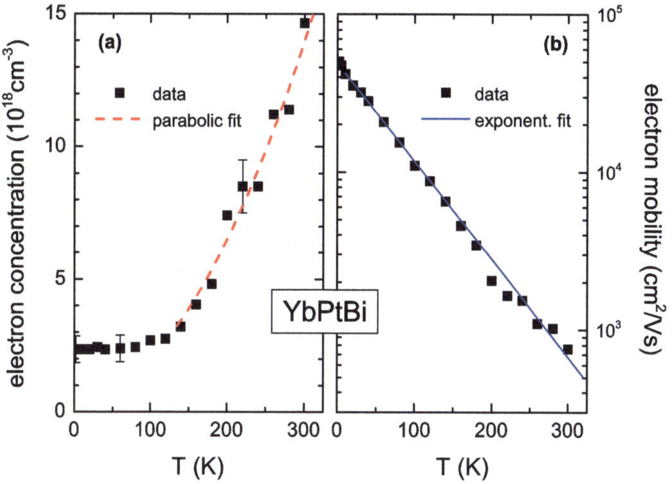

Fig. 4.10 Electron concentration (**a**) and mobility (**b**) in YbPtBi as obtained from Hall measurements with the correction from the optical-conductivity fits, as described in the text. The hole concentration and mobility are equal to 5.2×10^{20} cm^{-3} and $\sim$10 cm^2/Vs, respectively, and both are basically temperature independent. At higher temperatures, the electron concentration follows a T^2 behavior, as indicated by the dashed line, while the election mobility can nicely be fitted in the entire temperature range by an exponent shown as the solid line. Reproduced with permission from [146]. Copyright (2017) by the American Physical Society

4.3.3.2 Optics

Figure 4.11 displays the results of our optical investigations. The measured reflectivity spectra $R(\nu)$ are plotted in panel (a) for various temperatures. The Drude-like interband contributions below 2000 cm^{-1} are separated from the interband transitions at higher frequencies. The characteristic downturn in $R(\nu)$ near the plasma edge ($\sim$1000 cm^{-1}), however, reaches only 0.45, indicating some overlap of the intra- and interband contributions in the mid-infrared region.

Panel (c) of Fig. 4.11 shows the optical conductivity as derived from the Kramers-Kronig analysis of the reflectivity data. Here not only the absorption bands due to the intra- and interband transitions can be separated; it also becomes apparent that the low-frequency response consists of two Drude-like terms.

The dielectric constant $\varepsilon_1(\nu)$ is presented in Fig. 4.11b. It is negative below approximately 1000 cm^{-1}, due to the free-carrier (Drude) contributions and reveals a broad maximum at around 3500 cm^{-1} signaling an interband absorption edge at somewhat lower frequencies [75, 79] and confirming the assignment of the optical-conductivity modes above 2000 cm^{-1} to interband transitions. At $T < 100$ K, i.e. in the same temperature range, where $\rho_{\mathrm{dc}}(T)$ demonstrates the inflection point, a shoulder in $\varepsilon_1(\nu)$ appears at approximately 1600 cm^{-1}. The shoulder becomes more pronounced as the temperature decreases. It likely corresponds to an additional interband transition and hence indicates a possible band-structure modification at around $T \approx 90$ K.

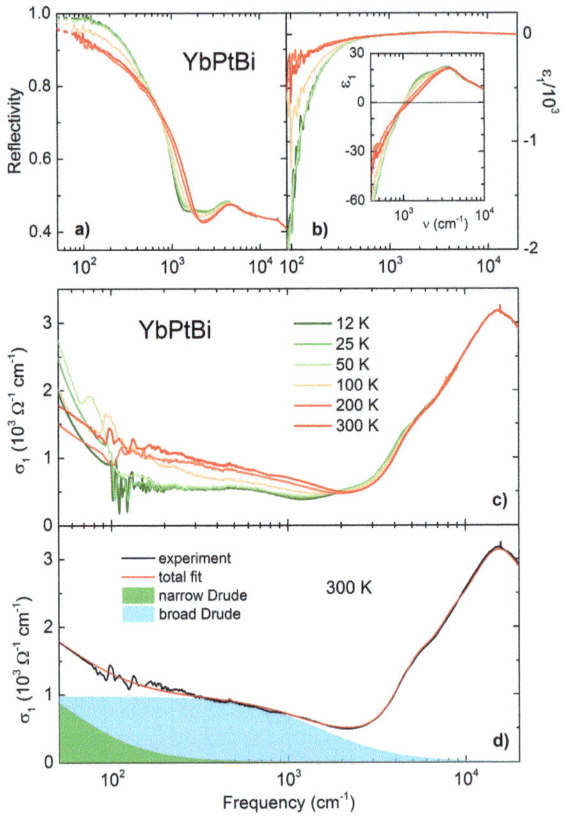

Fig. 4.11 Frequency-dependent reflectivity (**a**), dielectric constant (**b**), and optical conductivity (**c**) of YbPtBi for selected temperatures between 12 and 300 K. The inset in panel (**b**) zooms in $\varepsilon_1(\nu)$ near the zero crossing. Panel **d**: fit of the optical conductivity at 300 K using a Drude-Lorentz model. The intraband conductivity can be represented with two Drude components. The reflectivity and dielectric constant are fitted using the same Drude-Lorentz model (fits are not shown). Reproduced with permission from [146]. Copyright (2017) by the American Physical Society

Zero-line crossings of the $\varepsilon_1(\nu)$ curves (the inset of Fig. 4.11b) shift towards higher frequencies as T increases. This reflects the increased carrier concentration at higher temperatures: the zero-line crossings of $\varepsilon_1(\nu)$ can be taken as a measure of the screened plasma frequency, $\omega_{pl}^{scr}/2\pi$, which in turn is proportional to the carrier concentration.

In order to get more insight into the charge carrier dynamics, we simultaneously fit the experimental spectra of $\sigma_1(\nu)$, $\varepsilon_1(\nu)$, and $R(\nu)$ for each temperature by the Lorentz-Drude model. Results of these fits are exemplified in the panel (d) for σ_1 at $T = 300$ K. For the entire temperature range of our measurements, we were not able to describe the observed conductivity with a single Drude term, but obtained a satisfactory description only for two such contributions. These Drude terms have very different (by more than an order of magnitude) scattering rates, $1/\tau$. Analysis of Shubnikov–de Haas oscillations in YbPtBi yields the effective carrier masses m^* close to the free-electron mass m_e for any band, ranging from ~0.5 to 1.5 m_e [271]. Thus, the large difference between the electron and hole mobilities in YbPtBi should be mostly related to the difference in the scattering rates ($\mu = e\tau/m^*$) and one can ascribe the Drude term with the smaller scattering rate (the narrow Drude) to the highly mobile electrons and the term with larger scattering rate (the broad Drude) to the holes with low mobility.

The optical fits with the two Drude terms allow us to separate the contributions from the electrons and holes into the total conductivity. The bold dots in Fig. 4.7 correspond to the values obtained as $\nu \to 0$ extrapolations of the two Drude contributions. As one can see from the figure, their sum nicely follows the dc-conductivity curve. It is also apparent that at low temperatures electrons dominate the conductivity, while as $T \to 300$ K, the contributions of elections and holes to σ_{dc} become comparable, the electrons still providing a larger contribution.

4.3.3.3 Carrier Mobilities

Using the interpolations between the points, obtained from the optical-data fits, and the carrier concentrations, obtained from the Hall measurements, we calculate the carrier mobilities of electrons, μ_e, and holes, μ_h. The electron mobility is plotted in Fig. 4.10b. As the hole contribution to the dc conductivity is (much) lower than the electron contribution (Fig. 4.7), the values for electron motility do not differ much from the values, obtained using the one-carrier-type approximation. However, the hole mobility, calculated with the use of the optical fits, is much lower than the one obtained in the one-carrier-type approach. We find $\mu_h = (10 \pm 5)$ cm^2/Vs, being basically temperature independent. However, we cannot exclude a weak temperature dependence of μ_h within the given error bar: the hole scattering rate changes somewhat as a function of T.

As one can see from Fig. 4.10b, the electron mobility shoots up exponentially as T decreases.[1] We could fit μ_e with $\mu_e(T) = \mu_e(0)\exp(-T/T_0)$ in the entire temperature range. We have found $\mu_e(0) = 50\,000$ cm^2/Vs and $T_0 = 70$ K. The exponential behavior of μ_e is quite remarkable: usually, even in the materials with the highest reported mobilities, such as e.g. the Weyl semimetal NbP, $\mu(T)$ saturates at temperatures below some 20–30 K [17]. The exponential behavior of $\mu_e(T)$ in YbPtBi persists down to our lowest temperature (2.5 K) despite leveling off the electron concentration at $T < 100$ K, Fig. 4.10a. This signals collapsing of the electron scattering in YbPtBi at low temperatures. The electron concentration as a function of temperature does not show a sign of Arrhenius behavior at any T, the flat behavior at low temperatures being followed by a power law, $\mu_e \propto T^2$, at $T > 100$ K.

[1] We note that μ_e^{OCT} demonstrates the same exponential behavior as μ_e, with $\mu_e(0)$ and T_0 being only slightly changed.

4.3.4 Conclusions

From our transport, magnetotransport, and optical investigations of the half-Heusler compound YbPtBi at $T \geq 2.5$ K (i.e. not in the heavy-fermion state), we can separate two conduction channels caused by electrons and holes. Electrons posses a temperature-dependent concentration, high mobility, and low scattering rates. Holes form the second channel with the concentration and mobility that are basically temperature independent. While at high temperatures, i.e. 200–300 K, both channels provide comparable contributions to the dc transport, at low temperatures the electron channel dominates overwhelmingly. This is due to very high mobility, $\mu_e(T = 2.5\text{K}) = 50\,000\ \text{cm}^2/\text{Vs}$, and low scattering of the electrons at $T < 100$ K. To the best of our knowledge, the values of electron mobility in YbPtBi, are record high for half-Heusler compounds.

Chapter 5
Multifold Semimetals

An interesting example of the diverse world of topological semimetals is multifold semimetals. In these compounds, the topologically protected band crossings of degeneracy higher than two are described by Weyl-like Hamiltonians: linear in momentum and in effective spin, which can be larger than 1/2 [6, 11].

Recently, a number of compounds from the cubic space group 198 (SG198), which is non-centrosymmetric and has no mirror planes, were confirmed to possess such multifold crossing points [279–287]. In these materials, the band crossings with different chiralities are situated at different energy positions, providing thus a realization of electronic chiral crystals. Remarkably, the crystals of these compounds can be grown as single enantiomers, i.e., with a given crystalline and electronic chirality [285, 288].

Generally, optics seems to be an ideal tool for studying chiral materials and for manipulating the chiral degrees of freedom, as the circularly polarized light can be directly coupled to the chiral quasiparticles within such solids. For such investigations, it is essential to possess an advance knowledge of the linear optical response for the materials of interest. In other families of topological semimetals, optical spectroscopy, a genuine bulk-sensitive technique, was shown to provide an inside into the bulk electronic properties, see the previous Chapters and review papers [23, 24, 289]. Thus, efforts were recently taken to calculate and to measure the frequency-dependent conductivity of the chiral multifold compounds. In this Chapter, we present our results on two members of this family—RhSi and PdGa. These investigations are published in [290, 291].

A. V. Pronin, *Linear Electrodynamic Response of Topological Semimetals*,
Springer Series in Solid-State Sciences 199,
https://doi.org/10.1007/978-3-031-35637-7_5

5.1 RhSi

One of the peculiar effects, predicted for such electronic chiral crystals [292], is the quantized circular photogalvanic effect (QCPGE). In this non-linear optical phenomenon, the helical (i.e., circularly polarized) photons excite the chiral band carriers in such a way that the resultant photocurrent is quantized in units of material-independent fundamental constants. Recently, the observation of QCPGE has been reported in RhSi [293]—a member of SG198 and an established multifold semimetal [280–282].

These remarkable theoretical results and experimental observations demand further optical characterizations of RhSi, in particular since the knowledge of frequency-dependent conductivity is essential for proper interpretation of QCPGE experiments. Besides, optical conductivity of the multifold semimetals has been predicted to demonstrate a linear $\sigma_1(\omega)$ [97], similarly to (1.33). Further, optical conductivity has been specifically calculated for RhSi from its band structure [294]. Here, we experimentally examine these theoretical results.

5.1.1 *Experiment*

Single crystals of RhSi were grown in the same way as described in [293]. The vertical Bridgman crystal-growth technique was utilized to grow the crystals from the melt using a slightly off-stoichiometric composition (excess Si). First, a polycrystalline ingot was prepared using arc-melt technique by mixing the stoichiometric amount of constituent Rh and Si elements of 99.99% purity. Then the crushed powder was filled in a custom-designed sharp-edged alumina tube and finally sealed inside a tantalum tube with argon atmosphere. A critical composition with slightly excess Si was maintained to ensure a flux growth inside the Bridgman ampule. The whole assembly was heated to 1550 °C with a rate of 200 °C and halted for 12 h to ensure good mixing of the liquid. Then the crucible was slowly pulled to ~1100 °C with a rate of 0.8 mm/h and finally quenched to room temperature. The temperature profile was controlled by attaching a thermocouple at the bottom of the tantalum ampule containing the sample. Single crystals with average linear dimensions of a few millimeters were obtained. The crystals were first analyzed with a white beam backscattering Laue x-ray diffractometer at room temperature. The obtained single and sharp Laue spot could be indexed by a single pattern, revealing excellent quality of the grown crystals without any twinning or domains. The structural parameters were determined using a Rigaku AFC7 four-circle diffractometer with a Saturn 724+ CCD-detector applying graphite-monochromatized Mo-Kα radiation. The crystal structure was refined to be cubic $P2_13$ (SG198) with the lattice parameter $a = 4.6858(9)$ Å.

Temperature-dependent transport measurements (longitudinal dc resistivity and Hall) were performed in a custom-made setup at temperatures down to 2 K. The results of these measurements are displayed in Fig. 5.1d and e. A metallic behavior with a residual resistivity of 1.86×10^{-4} Ωcm was observed. The Hall measurements evidenced electron conduction with a moderate increase of carrier concentration upon increasing temperature.

Optical reflectivity, $R(\nu)$, was measured on a polished surface [290] of a RhSi single crystal (with roughly 1.5×1.5 mm^2 in lateral dimensions) over a broad frequency range from 80 to 20 000 cm^{-1} (10 meV–2.5 eV) at several different temperatures ($T = 10, 25, 50, 75, 100, 150, 200, 250, 295$ K). The spectra in the far-infrared (below 700 cm^{-1}) were collected with a Bruker IFS 113v Fourier-transform spectrometer using in situ gold coating of the sample surface for reference measurements. At higher frequencies, a Bruker Hyperion infrared microscope attached to a Bruker Vertex 80v FTIR spectrometer was used. For these measurements, freshly evaporated gold mirrors served as reference. No sample anisotropy was detected in agreement with the cubic crystallographic structure.

For Kramers-Kronig analysis, zero-frequency extrapolations have been made using the Hagen-Rubens relation in accordance with the temperature-dependent longitudinal dc resistivity measurements. For high-frequency extrapolations, we utilized the x-ray atomic scattering functions [145] followed by the free-electron behavior, $R(\omega) \propto 1/\omega^4$, above 30 keV.

We found that the skin depth of the probing radiation exceeds 30 nm for all temperatures and frequencies (in the far-infrared range, it is above 200 nm). Hence, our optical measurements reflect the bulk properties of RhSi.

5.1.2 Results and Discussion

Figure 5.1 displays the optical reflectivity $R(\nu)$ and the real parts of the optical conductivity and dielectric constant for the studied RhSi sample in the full measurement-frequency range at three different temperatures. The overall temperature evolution of the spectra is minor. For frequencies higher than ~8 000 cm^{-1} (~1 eV), the optical properties are fully independent of temperature. For the spectra analysis, we performed standard Drude-Lorentz fits [75], where the Drude terms describe the free-carrier response, while the Lorentzians mimic the interband optical transitions and phonons. Examples of such fits are presented in Fig. 5.2. Here we kept the zero-frequency limit of optical conductivity to be equal to the measured dc-conductivity value at every temperature. No other restrictions on the fit parameters were imposed. For best possible model description, the experimental spectra of $R(\nu)$, $\sigma_1(\nu)$ and $\varepsilon_1(\nu)$ were fitted simultaneously.

5.1.2.1 Electronic Response

At low frequencies, RhSi demonstrates a typical (semi)metallic response. The intraband contribution to the spectra can be best fitted with two Drude components, which have different scattering rates. Such multi-component Drude fits are often used to

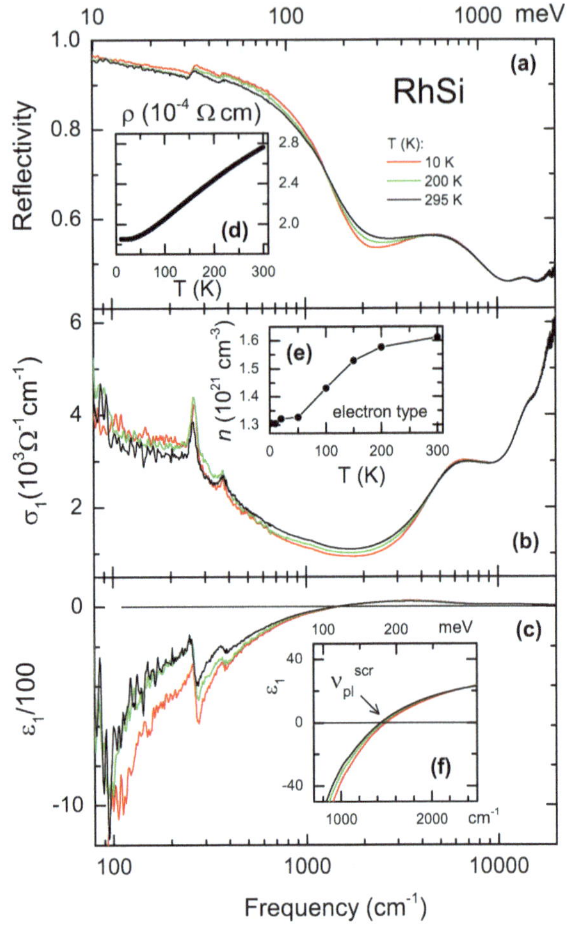

Fig. 5.1 Optical reflectivity (**a**) and the real parts of the optical conductivity (**b**) and the dielectric permittivity (**c**) of RhSi at $T = 10$, 200, and 295 K. Note logarithmic x-scale. The insets show: the dc resistivity vs. T (**d**), the Hall electron concentration vs. T (**e**), and a zoom of the permittivity spectra near the zero crossings (**f**). Reprinted from [290] under CC-BY 4.0 licence. Copyright (2020) by the authors

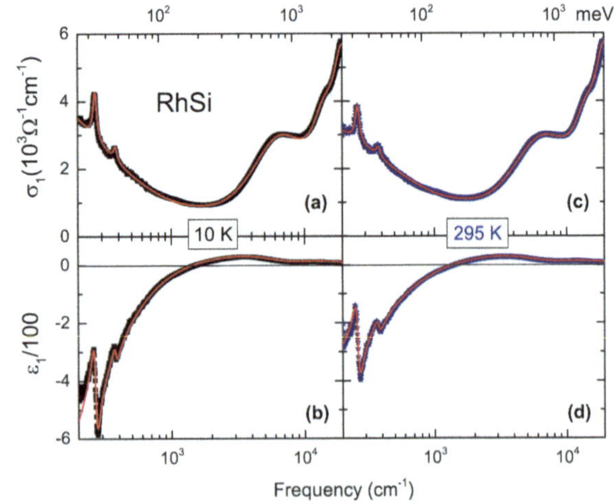

Fig. 5.2 Drude-Lorentz fits (lines) of the measured optical spectra (symbols) of σ_1 and ε_1 of RhSi at 10 and 295 K, as indicated. Note logarithmic x-scale. Reprinted from [290] under CC-BY 4.0 licence. Copyright (2020) by the authors

describe the optical response of multiband systems, particularly different semimetals, see the previous chapters and [231, 295]. In the case of RhSi, the two-Drude approach can be justified by the presence of a few bands crossing the Fermi level [69, 280, 281, 294], see Fig. 5.4a: one set of bands is around the R point and the others are near the Γ and M points of the BZ. The first set provides the dominating contribution to the free-carrier response and is also responsible for the electron type of conduction. Still, the second Drude term is necessary to describe the optical spectra accurately. Let us note that exact interpretation of such a two-Drude approach is arguable.

The scattering rates of the Drude terms are found to be 250 cm^{-1} ($\sim$30 meV) and 800 cm^{-1} (100 meV) at 100 K. These values correspond to the mid-ranges for the spectra at all measurement temperatures. At T different from 100 K, the scattering rates change only within $\pm$20% of these values. The mid-range relaxation times are, hence, 21 and 6.6 fs. These values are somewhat larger than the value reported in another optical study of RhSi [293] indicating the improved quality of RhSi samples investigated in this work.

At around 1500 cm^{-1} ($\sim$200 meV), a characteristic plasma edge is observed in $R(\nu)$. This edge correlates with the zero crossing of $\varepsilon(\nu)$, which corresponds to the screened plasma frequency, $\nu_{pl}^{scr} = \nu_{pl}/\sqrt{\varepsilon_{\text{inf}}}$. Here, $\varepsilon_{\text{inf}} = 65 \pm 5$, is the cumulative contribution of the higher-frequency optical transitions to ε_1 as obtained from the fits and ν_{pl} is the unscreened plasma frequency. As best seen from Fig. 5.1f, the screened plasma frequency is $\nu_{pl}^{scr} = (1470 \pm 30)$ cm^{-1} [$\hbar\omega_{pl}^{scr} = (182 \pm 4)$ meV] and independent of temperature, the corresponding unscreened plasma frequency being $\nu_{pl} = (11\,900 \pm 700)$ cm^{-1} [$\hbar\omega_{pl} = (1470 \pm 90)$ meV]. This value of plasma frequency coincides within the experimental uncertainty with the value obtained from the Drude fits, $\nu_{pl} = (11\,200 \pm 600)$ cm^{-1} [$\hbar\omega_{pl} = (1390 \pm 80)$ meV], which includes contributions from both Drude terms and also shows no T-dependence.

The absence of any detectable temperature dependence of ω_{pl} is in qualitative agreement with the very modest temperature-induced change of the carrier concentration: $n(T)$ increases by only 20% as T goes from 2 to 300 K, see Fig. 5.1e.

The relatively large value of ω_{pl} (cf. the results for other nodal semimetals [170, 172, 235, 236, 258]) and the fairly high ($\sim 10^{21} cm^{-3}$) free-electron density of RhSi, see Fig. 5.1d, are consistent with the results of band structure calculations [69, 280, 281, 294], which show that the Fermi level in RhSi is quite deep in the conduction band for the electron momenta near the corners (R points) of the BZ (Fig. 5.4a). This situation makes the optical response of RhSi similar to the one observed by us in the Dirac semimetal Au_2Pb: free carriers dominate the low-frequency ($\nu < 2000$ cm^{-1}, $\hbar\omega < 250$ meV) region of $\sigma_1(\nu)$ and $\varepsilon(\nu)$. Still, unlike the situation in Au_2Pb, the optical transitions between the linearly dispersing bands can be resolved in RhSi.

In Fig. 5.3 our optical findings at $T = 10$ K are plotted together with the results of band-structure-based calculations from [294] and with the previously reported measurements of [293]. The experimental curves follow each other quite well. The deviations between the curves can be explained by different free-carrier contributions (cf. the difference in the scattering times discussed above) and probably by a somewhat more accurate Kramers-Kronig analysis utilized in the present work: our

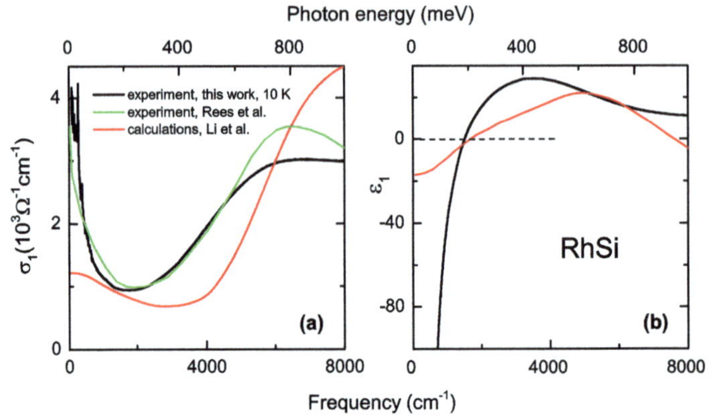

Fig. 5.3 Comparison of the experimental and calculated optical spectra of RhSi: optical conductivity (**a**) and dielectric function (**b**). The literature data are by Rees at al. [293] and Li et al. [294]. Reprinted from [290] under CC-BY 4.0 licence. Copyright (2020) by the authors

reflectivity measurements are performed in a broader frequency range as compared to the measurements from [293].

Despite some discrepancy between the calculations and both experimental curves in Fig. 5.3a, the match can be considered as satisfactory. One has to keep in mind that calculations of the optical conductivity from the electronic band structure are rather challenging, particularly for semimetals: a survey of the available literature reveals only a qualitative match between the calculated optical conductivity and experimental results for a wide range of nodal semimetals studied recently [174, 220, 231, 236, 237]. Nevertheless, both the low-energy features of the interband experimental $\sigma(\nu)$—the initial (i.e., for the frequencies just above the Drude roll-off) linear increase and the further flattening—are reproduced by theory.

In order to establish a better connection between the features observed in the most interesting, low-energy part, of the experimental conductivity and the interband optical transitions, we show our $\sigma(\nu)$ together with the low-energy band structure of RhSi in Fig. 5.4. Additionally, we plot the interband contribution to the optical conductivity, $\sigma_1^{\mathrm{IB}}(\nu)$, obtained by subtracting the Drude fits and the sharp phonon peaks (discussed below) from the measured spectra. The interband optical conductivity found this way is pretty much linear in frequency for $\nu < 3000\ \mathrm{cm}^{-1}$. We found that this approximate linearity is robust: varying the fit parameters within the uncertainty, set by the experimental spectra, does not compromise it [290].

At the lowest frequencies (below approximately 2500 cm^{-1} or 0.3 eV), the interband conductivity is entirely due to transitions in the vicinity of the Γ point. No other interband optical transitions are possible (either the direct gap between the bands is too large, or the transitions are Pauli blocked). The bands near the Γ point are all roughly linear (two of them are basically flat), thus a linear-in-frequency interband conductivity is expected [97]. Indeed, as already noticed, $\sigma_1^{\mathrm{IB}}(\nu)$ is proportional to frequency in this range (marked with the orange arrows). At somewhat higher ν

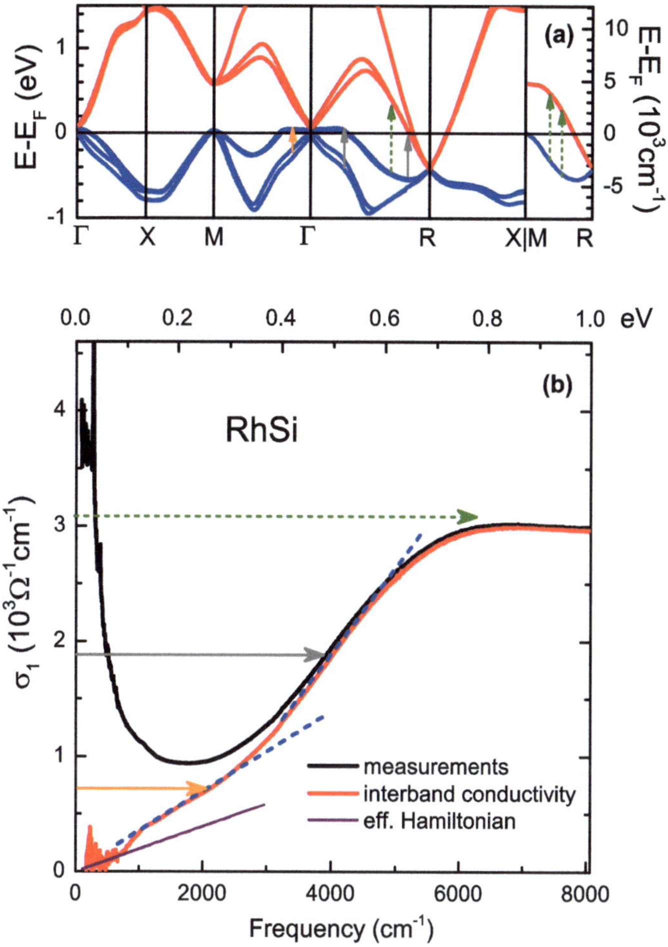

Fig. 5.4 Low-energy electronic structure of RhSi (**a**) and its optical conductivity (**b**). The electronic structure is calculated within the Topological Materials Database project [69]. Note that the band color code follows this reference and is not always related to the band position relative to the Fermi level. Both, the as-measured conductivity spectra and the spectra after subtraction of the free-carrier response, $\sigma_1^{\mathrm{IB}}(\nu)$, are shown in (**b**). The optical transitions responsible for the characteristic features in $\sigma_1^{\mathrm{IB}}(\nu)$ are depicted as vertical arrows in (**a**). The corresponding frequency scales are indicated by the horizontal arrows of the same color in (**b**). The solid orange arrow represents the transitions between the approximately linear bands (including the almost flat bands) near the Γ point and the corresponding linear $\sigma_1^{\mathrm{IB}}(\nu)$. The solid grey arrows indicate two different processes coinciding in frequency: the onset of the downturn of the flat bands and the onset of interband transitions in the vicinity of the R point. These processes lead to additional contributions to $\sigma_1(\nu)$. The dashed green arrows correspond to the transitions between the almost parallel bands along the M–R line, leading to a maximum in $\sigma_1(\nu)$. Transitions with approximately, but not exactly, the same energy (e.g., along the Γ–R line) are responsible for smearing out this maximum and are indicated with the same arrow. The dashed blue lines is a guide to the eye. The solid purple line is an extrapolation of the effective-Hamiltonian computations from [97] to higher frequencies. Reprinted from [290] under CC-BY 4.0 licence. Copyright (2020) by the authors

($\sim$3000$-$4000 cm^{-1}, 0.4$-$0.5 eV), the flat bands start to disperse downward, thus the linearity of $\sigma_1^{\text{IB}}(\nu)$ is not expected anymore. However, the interband contributions in the vicinity of the R points become allowed at roughly the same energy (cf. the two grey arrows in panel (a)). These transitions provide a dominating contribution to conductivity, and the linear-in-frequency increase of $\sigma_1^{\text{IB}}(\nu)$ is restored with a larger slope (the grey arrow in panel (b)). At $\nu \geq 6000$ cm^{-1} (0.8 eV), the optical conductivity flattens out, forming a broad flat maximum. We attribute it to the transitions between the almost parallel bands along the M–R line, shown as the dotted green arrows. The maximum is not sharp because there are many other transitions with comparable frequencies—see, e.g., the dotted green arrow between the Γ and R points. After the relatively flat region, $\sigma_1^{\text{IB}}(\nu)$ continues to rise (see Fig. 5.1b), as more and more bands get involved in optical transitions.

In Fig. 5.4b, we also compare our results with the effective-Hamiltonian calculations [97] for the contributions near the Γ point. An extrapolation of these calculations, originally performed for $\nu < 320$ cm^{-1} (40 meV), to higher frequencies is shown as a solid purple line. The experimental $\sigma_1^{\text{IB}}(\nu)$ is generally steeper than the results of these calculations. A very similar behavior of the experimental conductivity versus such effective-Hamiltonian calculations has also been reported in [293]. Perhaps, at the lowest frequencies the match between the experiment and the model is better, but our signal-to-noise ratio is not sufficient for final conclusions (one should also remember that the $\sigma_1^{\text{IB}}(\nu)$ spectrum is obtained utilizing a Drude-terms subtracting procedure). In any case, the mismatch can be related to deviations of the bands from linearity even at low energies [69, 280, 281, 294]. This can be clarified in more advanced band-structure-based optical-conductivity calculations, which are beyond the scope of our study [296].

Having established the connection between the features in the experimental low-energy interband conductivity and the band structure, we would like to add another note on the intraband response. The exact shape of the free-carrier contribution is obviously sample dependent, because, e.g., the impurity scattering rate differ from sample to sample. However, if the total amount of doping is low enough, the plasma frequency is fixed by the position of the Fermi level, which, in turn, can be found within the band structure calculation procedure. Such calculations [294] produced $\hbar\omega_{pl} = 1.344$ eV, in excellent agrement with our result obtained above, 1.39 ± 0.08 eV. Furthermore, the calculated and the observed spectral positions of the screened plasma frequency (the zero-crossing points of ε_1) also match very well each other, as can be seen in Fig. 5.3b. This agreement of the plasma frequencies also indicates a good quality of our sample in terms of low defect and impurity concentration.

5.1.2.2 Phonons

Based on its crystallographic symmetry, RhSi is supposed to show five infrared-active phonon modes; however, we can clearly identify only two of them in our spectra. The other modes are likely too weak to be resolved on top of the electronic background

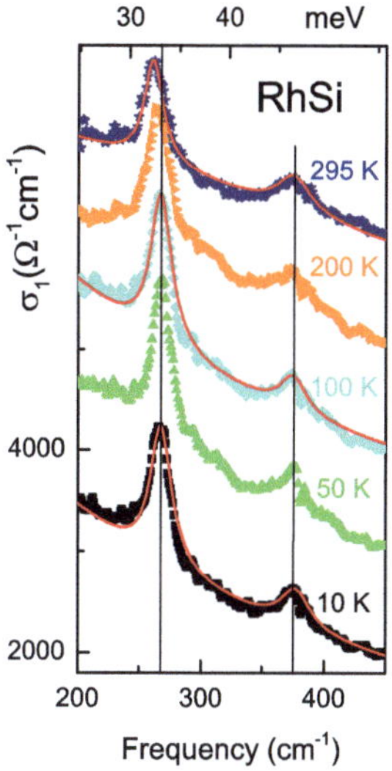

Fig. 5.5 Optical conductivity of RhSi at the frequencies near the observed phonon modes for selected temperatures. The y-scale corresponds to the 10-K spectra. The spectra for other temperatures are shifted upwards for clarity. Results of Lorentzian fits of the phonon modes (on an electronic background) are shown for 10, 100 and 295 K as solid red lines. The black vertical lines are a guide to the eye. Reprinted from [290] under CC-BY 4.0 licence. Copyright (2020) by the authors

within the available experimental accuracy. The two sharp phonon modes fall in the spectral range from 200 to 400 cm^{-1} (25–50 meV), see Fig. 5.5c. The positions of these modes at $T = 10$ K are marked with thin vertical lines.

Both modes can be accurately described by Lorentzians at any temperature. No asymmetric (Fano-like) models are necessary. This is in contrast to the situation in FeSi—an isostructural analogue of RhSi with presumably important role of electron correlations—where strong phonon-line asymmetry was reported in the optical spectra and related to electron-phonon coupling [297]. The absence of detectable electron-phonon coupling indicates that phonon-mediated electron-correlation effects are likely of no relevance in RhSi.

Let us finally mention that the phonon modes observed in RhSi demonstrate a usual broadening as temperature rises. Additionally, the low-frequency mode shows a small softening of its central frequency with increasing T, which can be explained by usual thermal expansion (possible softening of another mode might be not resolved because of a larger width of this mode).

5.1.3 Conclusions

We have measured the broadband optical response of the multifold semimetal RhSi. Infrared-active phonons and electronic transitions are revealed in this study. The phonon modes demonstrate a trivial temperature dependence with no indications of strong electron-phonon coupling. The intraband electronic contribution (Drude) is relatively strong with the (unscreened) plasma frequency of approximately 1.4 eV. The interband optical conductivity demonstrates a linear increase at low frequencies (below 300 meV). We interpret this increase as a signature of the transitions between the linear bands crossing around the Γ point. At somewhat higher frequencies (400–600 meV), contributions from the linear bands near the R point (Pauli-blocked at lower frequencies) manifest themselves as an increased slope of $\sigma_1(\omega)$. These observations confirm the predictions for the optical-conductivity behavior in multifold semimetals.

5.2 PdGa

5.2.1 Experiment

PdGa single crystals were grown from its melt by self-flux technique, as described in [285, 288]. First, a polycrystalline ingot was prepared using the arc melt technique with a stoichiometric mixture of high-purity Pd and Ga. Then, the crushed powder was filled in a thin-wall alumina crucible and finally sealed in a quartz ampoule. The crystal growth was done under partial vacuum of 3 mbar. The ampoules were heated to 1100 °C, halted there for 12 h, and then slowly cooled to 900 °C with a rate of 1.5 °C/h. Finally, the samples were cooled to 800 °C with a rate of 50 °C/h, annealed for 120 h and then cooled to 500 °C with a rate of 5 °C/h. PdGa single crystals with average linear dimensions of a few mm were obtained.

The crystals were first analyzed with a white beam backscattering Laue x-ray diffractometer at room temperature. The obtained single and sharp Laue spot could be indexed by a single pattern, revealing excellent quality of the grown single-enantiomer crystals without any twinning or domains, see [291] for a Laue-pattern example. The structural parameters were determined using a Rigaku AFC7 four-circle diffractometer with a Saturn 724+ CCD-detector applying graphite-monochromatized Mo-Kα radiation. The crystal structure was refined to be cubic $P2_13$ (SG198) with the lattice constant $a = 4.896$ Å.

Temperature-dependent transport measurements (longitudinal dc resistivity and Hall) were performed in custom-made setups at temperatures down to 2 K.

Temperature-dependent ($T = 10$–295 K) optical reflectivity, $R(\nu)$, was measured on a PdGa single crystal (with roughly 1.5×1.5 mm^2 in lateral dimensions) over a broad frequency range from $\nu = \omega/(2\pi) = 100$ to $20\,000$ cm^{-1} (12 meV–3 eV). The spectra in the far-infrared (below 700 cm^{-1}) were collected with a Bruker IFS 113v

Fourier-transform spectrometer using in situ freshly evaporated gold over-filming technique for reference measurements. At higher frequencies, a Bruker Hyperion infrared microscope attached to a Bruker Vertex 80v FTIR spectrometer was used. For these measurements, freshly evaporated gold mirrors on glass substrates served as reference. No sample anisotropy was detected, which is in agreement with the cubic crystallographic structure.

For the Kramers-Kronig analysis, the zero-frequency extrapolations were made using the Hagen-Rubens relation in accordance with the temperature-dependent longitudinal dc resistivity measurements. For high-frequency extrapolations, we utilized the x-ray atomic scattering functions [145] followed by the free-electron behavior, $R(\omega) \propto 1/\omega^4$, above 30 keV.

5.2.2 *Calculations*

The band structure and optical properties of PdGa were calculated by first-principles calculations based on the density-functional theory with the Perdew-Burke-Ernzerhof exchange-correlation functional implemented in Quantum ESPRESSO [298]. Norm-conserving pseudopotentials with the generalized gradient approximation (GGA) are adopted in this work. We used the experimental structural parameters [288] without any geometry optimization. The energy cutoff of $35.0R_y$ and the k-point grid with $24 \times 24 \times 24$ ($40 \times 40 \times 40$) points were adopted for the band-structure (optical-properties) calculations. The SOC effects as well as a frequency-independent broadening for interband transitions due to electron scattering (0.2 eV) were included in the calculations.

The bulk Fermi surfaces of PdGa (and of RhSi, for comparison) were calculated using WIEN2k's [299] full-potential linearized augmented plane wave methods with the Perdew-Burke- Ernzerhof exchange-correlation functional on a $32 \times 32 \times 32$ k-mesh, with account for SOC and with the lattice parameters taken from [288]. More information on the relevant computations can be found in [300].

5.2.3 *Results and Discussion*

Examples of the frequency-dependent optical spectra are shown in Fig. 5.6 for selected temperatures. The raw reflectivity shows typical metallic behavior with $R(\nu)$ approaching the unity as frequency diminishes, see panel (a). The spectra of the real part of optical conductivity [panel (b)] demonstrate a corresponding behavior: σ_1 increases as $\nu \to 0$. This is in qualitative agreement with a simple free-electron Drude model. The screened plasma frequency of free electrons, ν_{pl}^{scr}, can be estimated from the zero crossings of the permittivity spectra shown in panel (d). We found ν_{pl}^{scr} to be temperature independent at $\nu_{pl}^{scr} \simeq 11\,000$ cm^{-1}, corresponding to $\hbar\omega_{pl}^{scr} = 1.37$ eV.

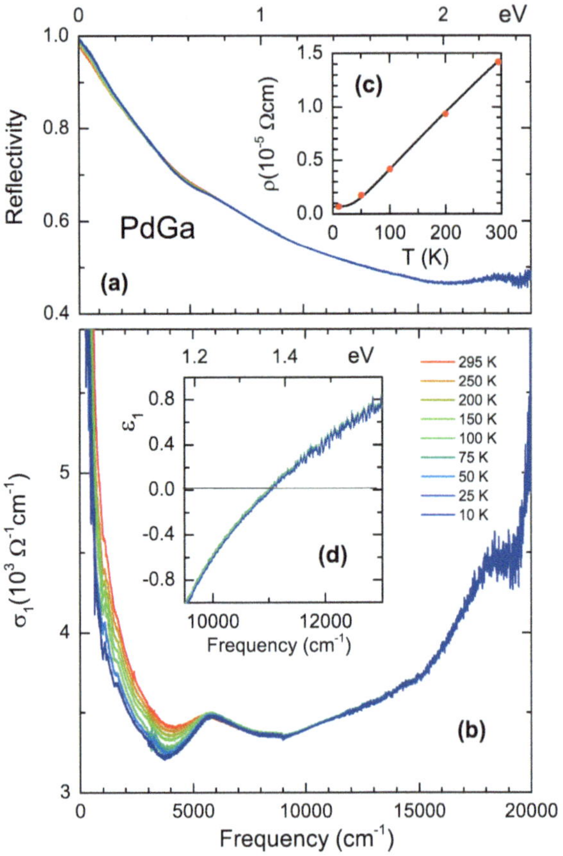

Fig. 5.6 PdGa optical reflectivity (**a**) and the real part of optical conductivity (**b**) at selected temperatures as indicated. The insets (**c**) and (**d**) show, respectively: the dc resistivity versus T (line) together with the inverse optical conductivity in the $\nu \to 0$ limit used in the fits of Fig. 5.2 (bold dots); and zoomed permittivity spectra near the zero-line crossing. Note that in panels (**a**), (**b**), and (**d**), the experimental curves are presented for all indicated temperatures. Reproduced with permission from [291]. Copyright (2021) by the American Physical Society

Comparing these findings with the experimental results on ν_{pl}^{scr} in many other nodal semimetals [170, 172, 235–237, 258, 290, 301, 302], one can immediately notice a very large value of ν_{pl}^{scr} in PdGa. For example, it is roughly one order of magnitude larger, than in the sister compounds, the multifold semimetals RhSi, $\nu_{pl}^{scr} \sim 1500$–$1700\ \text{cm}^{-1}$ [see above and [303]], and CoSi, $\nu_{pl}^{scr} \sim 600$–$800\ \text{cm}^{-1}$ [301, 302]. This "metallicity" of PdGa is obviously related to its band structure with the chemical potential situated far from the nodes, as discussed below. Our Hall measurements also reveal the metallic nature of PdGa: its electron density is high and almost temperature-independent, $n = (3 \pm 1) \times 10^{22}\ \text{cm}^{-3}$.

As seen in Fig. 5.6, the overall temperature evolution of the conductivity spectra is rather weak: only the low-energy free-electron part shows detectable T-induced changes due to the temperature-dependent electron scattering, in agreement with the metal-like optical response. At $\nu > 4000\ \text{cm}^{-1}$, the interband transitions start to become visible in $\sigma_1(\nu)$.

To analyze the optical spectra in a more quantitative way, we first performed a standard Drude-Lorentz fit [75] for a number of temperatures. The Drude contribution

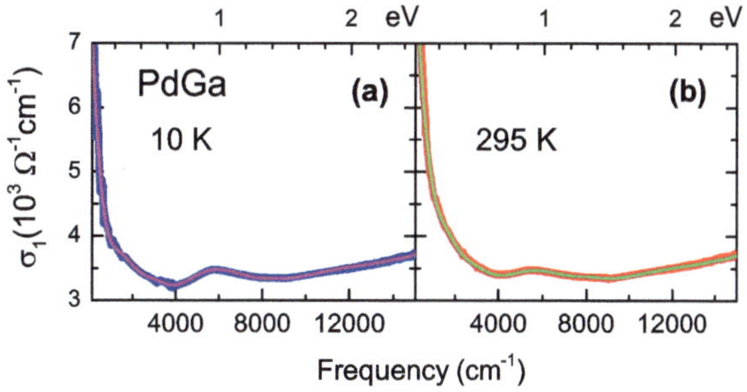

Fig. 5.7 Drude-Lorentz fits (lines) of the measured optical conductivity spectra (symbols) at 10 and 295 K. Reproduced with permission from [291]. Copyright (2021) by the American Physical Society

describes the intraband response, while the Lorentzians are used to fit the interband optical transitions. Examples of such fits are presented in Fig. 5.7. In all the fits, we kept the zero-frequency limit of optical conductivity equal to the measured dc-conductivity value at every temperature, see Fig. 5.6c. No other restrictions on the fit parameters were imposed. The fits were obtained by simultaneous fitting of $R(\nu)$, $\sigma_1(\nu)$ and $\varepsilon_1(\nu)$.

We found that we need at least two Drude components ("narrow" and "broad") with different scattering rates to provide accurate Drude-Lorentz fits to the experimental spectra. This approach is often used to describe the intraband optical response in different multiband materials [146, 173, 174, 204, 290, 295, 303–305], but the exact interpretation of the two Drude terms remains arguable [173, 290, 303]. PdGa possesses not two, but many different bands crossing the Fermi level, see Figs. 5.9 and 5.10. Hence, the two components might be associated, e.g., with two different *sets* of bands, or with different scattering mechanisms. In any case, the two-Drude approach utilized here should be considered just as a minimalist model to fit the intraband optical response in a Kramers-Kronig consistent way and to extract the total spectral weight (the plasma frequency) of itinerant carriers.

The fit parameters of the Drude terms are shown in Fig. 5.8 as functions of temperature. Because of the temperature-induced redistribution of the spectral weight between the terms, neither of the scattering rates (γ) is expected to follow the dc-resistivity temperature dependence accurately. Nevertheless, it primarily the temperature variation of the narrow-Drude scattering rate, which provides the reconciliation of the relatively strong temperature dependence of dc conductivity [the residual resistivity ratio is around 20, see Fig. 5.6c] and the fairly weak temperature evolution of intraband optical conductivity.

While the parameters of the Drude terms vary appreciably with temperature, the total intraband unscreened plasma frequency ν_{pl} remains almost temperature independent at $\sim$21 800 cm^{-1} ($\hbar\omega_{pl} \approx 2.7$ eV). This value is significantly larger than the one reported for RhSi ($\sim$11 300 cm^{-1} or 1.4 eV) [290], that is in qualitative agrement with the smaller bulk Fermi surface of RhSi, see Fig. 5.10. Also, the fit-based ν_{pl} in PdGa is consistent with the screened plasma frequency ν_{pl}^{scr}, obtained from zero crossings of $\varepsilon_1(\nu)$, if the higher-frequency dielectric constant ε_∞ is assumed to be around 1.4, which is a reasonable value [cf. Fig. 5.6d]. We recall that $\nu_{pl}^{scr} = \nu_{pl}/\sqrt{\varepsilon_\infty}$.

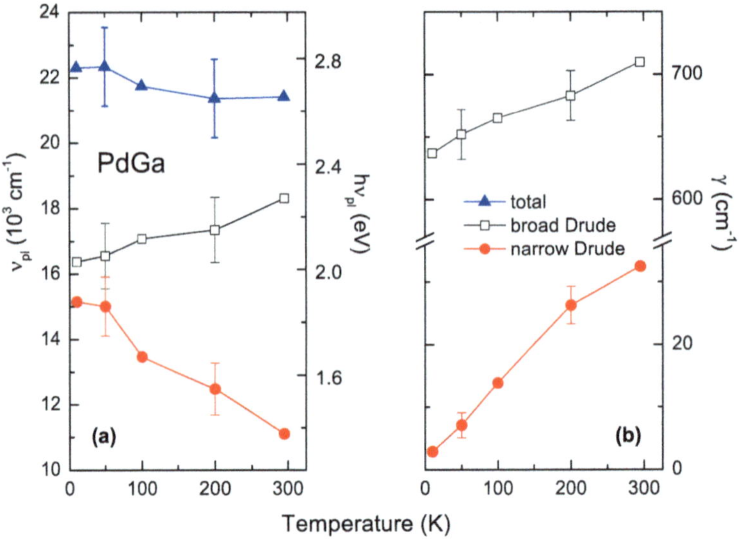

Fig. 5.8 Temperature-dependent parameters of the two Drude terms (narrow and broad) used in the Drude-Lorentz fits. Note different vertical scales for the plasma frequencies [panel (a)] and the scattering rates [panel (b)]. Reproduced with permission from [291]. Copyright (2021) by the American Physical Society

Fig. 5.9 Low-energy electronic band structure of PdGa with the spin-orbit coupling included. The multifold fermions are supposed to exists near the Γ and R points. Reproduced with permission from [291]. Copyright (2021) by the American Physical Society

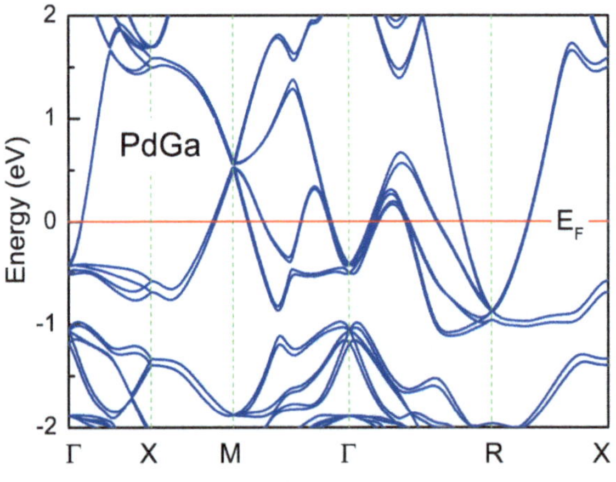

For a more elaborative analysis, we performed band-structure calculations for PdGa and then computed its interband optical conductivity, as described above. The results of the band-structure calculations are shown in Fig. 5.9. It is apparent that the Fermi level E_F crosses a number of electron- and hole-like bands and that the multifold nodes near the Γ and R points are situated quite deep (0.5 eV or more) below E_F. Hence, PdGa possesses an extended bulk Fermi surface, which occupies a significant portion of the BZ, see Fig. 5.10.

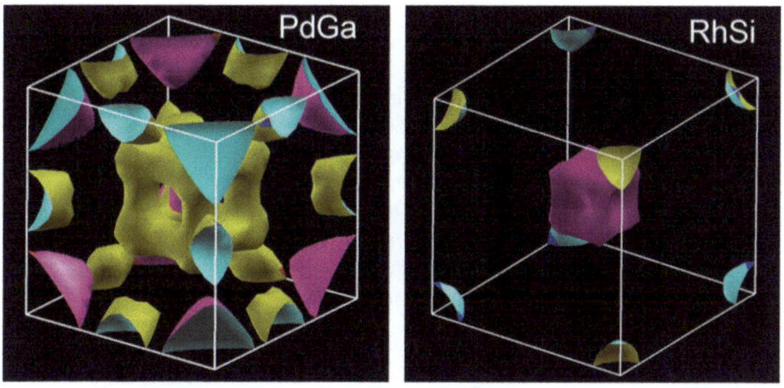

Fig. 5.10 Calculated bulk Fermi surface of PdGa (left) and its cut at the middle of the Brillouin zone (right). The k_z direction is perpendicular to the picture plane. Reproduced with permission from [291]. Copyright (2021) by the American Physical Society

The large bulk Fermi surface and the consequent intense intraband optical response make a direct comparison between the interband conductivity computed from the band structure and the experimental spectra challenging. As seen from Fig. 5.11, the interband portion of optical conductivity is not zero down to very low frequency. This might look surprising, as usually there is an onset (a Pauli edge) in the interband-conductivity spectra for the systems with the Fermi level situated far from the nodes, which is the case for PdGa. However, PdGa possess multiple bands with small energy separation, which (the bands) cross the Fermi level, see Fig. 5.9 and the inset of Fig. 5.11. The cumulative effect of the transitions between these bands causes the quasi-linearity of the interband $\sigma_1(\nu)$ in PdGa at low frequencies. We note that at $\nu \to 0$, $\sigma_1(\nu)$ flattens out. More details on band-selective optical transitions in PdGa can be found in [300].

In order to compare the experimental and computed interband conductivity, the intraband (Drude) response can be subtracted from the former spectra, as it was done, e.g., in [231, 290, 303, 306]. This procedure may obviously produce ambiguities in determining the interband portion of the experimental $\sigma_1(\nu)$, especially at the lowest frequencies, where the interband conductivity is low. Hence, we followed a slightly different approach: instead of subtracting the Drude terms from the experimental spectra, we added them to the calculated interband conductivity. Basically, we performed a sort of fit with two Drude terms and the interband $\sigma_1(\nu)$ obtained from the band structure with a frequency-independent electron scattering rate of 0.2 eV ($\sim$ 1600 cm^{-1}). The results of this analysis are shown in Fig. 5.11. As a starting point, we used the Drude terms, obtained from our Drude-Lorentz fits (Fig. 5.7). For the best possible description of the experimental spectra, we had to slightly change the parameters of these terms, but the zero-frequency limit of $\sigma_1(\nu)$ remained to be equal to the inverse of the measured dc resistivity.

Having in mind that the band-structure-based computations of $\sigma_1(\nu)$ generally reproduce the experimental findings only qualitatively [172, 231, 236, 258, 289, 290], the match between theory and experiment can be considered as rather good: in

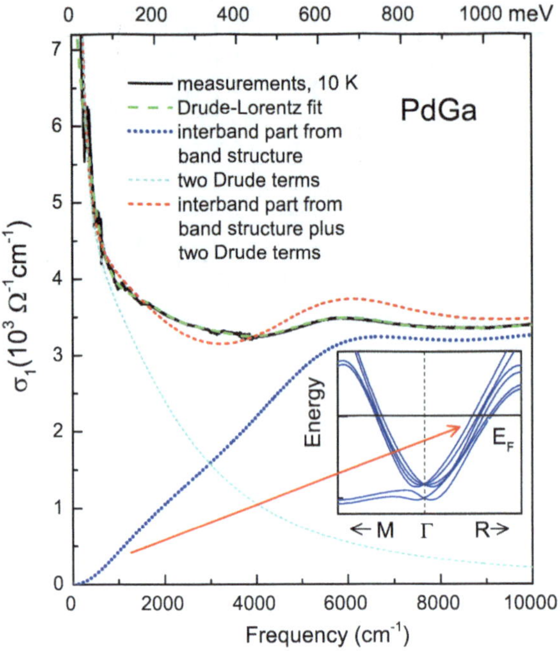

Fig. 5.11 Comparison of the measured (solid line) and calculated (dashed and dotted lines) optical conductivity in PdGa. The interband portion of $\sigma_1(\omega)$ (blue dotted line) is calculated from the band structure. Adding two Drude terms (cyan dashed line) to this curve provides a good qualitative match (red dashed line) to the experimental spectrum. The green dashed line is the fit from Fig. 5.2. The quasi-linear interband conductivity at low frequencies is due to the transitions between the multiple bands in the vicinity of the Fermi level, as shown in the inset. Reproduced with permission from [291]. Copyright (2021) by the American Physical Society

the present case the general conductivity level observed in experiment is reproduced by computations and the major feature of the interband $\sigma_1(\nu)$—the flat maximum at around 5500 cm^{-1}—is seen in both, computed and measured spectra. The remaining discrepancies can be attributed, e.g., to a frequency-dependent electron scattering in the investigated sample (as noticed above, we assumed a frequency-independent scattering rate in our computations).

Due to the very high free-carrier concentration and relatively large electron scattering (the broad Drude component), direct experimental verification of the linear interband optical conductivity at low energies, as it is predicted by theory for multifold semimetals with the nodes situated in the vicinity of E_F [97], is impossible for PdGa. Nevertheless, our experimental spectra can be well described as a sum of the interband conductivity, obtained from band structure, and a strong Drude-like free-carrier contribution. Let us also note that this strong electronic response prevents observation of any phonon modes on the top of it (based on its crystallographic symmetry, PdGa is supposed to have five infrared-active phonons).

5.2.4 Conclusions

We have studied the broadband optical conductivity of the multifold semimetal PdGa. A prominent metallic response is detected. The free-carrier Drude-like contribution with a temperature-independent plasma frequency ($\hbar\omega_{pl}^{scr} = 1.37$ eV) dominates the spectra at frequencies below 4000 cm^{-1}, preventing direct detection of the linear-in-frequency interband conductivity predicted for the multifold semimetals. At higher frequencies, the spectra calculated from the band structure reproduce the experimental spectra. Namely, the general conductivity levels obtained in experiment and in computations match each other and the frequency position of the most prominent feature in the experimental interband conductivity—the maximum at around 5500 cm^{-1} (680 meV)—is reproduced by the computations.

Chapter 6
Summary

When graphene shifted in the focus of condensed matter physics, its constant in frequency conductivity appeared as a peculiarity at first glance. Soon it became clear that this was the tip of an iceberg: materials with interesting band structure and topology are predicted and discovered at a rapid pace since. The Dirac cones in graphene are just a particular case of a class of systems that is not restricted to two dimensions. Different nodal semimetals, possessing conical bands in their bulk, are currently at the center of studies, and conductivity of these materials is directly related to the electronic band dispersion and dimensionality. Albeit ARPES is surely the most proper method to study the band structure and Fermi surface, its severe restriction to the sample surface often causes problems that can be overcome by optical methods, as a genuine bulk sensitive technique. In this monograph, we considered a representative selection of our recent experimental results on optical studies of different nodal semimetals. The presented examples demonstrate the abilities, as well as limitations, of linear optics to reveal the bulk band structure at low energies and to confirm the presence of the chiral electronic bands in some of the topological materials.

A. V. Pronin, *Linear Electrodynamic Response of Topological Semimetals*,
Springer Series in Solid-State Sciences 199,
https://doi.org/10.1007/978-3-031-35637-7_6

Appendix A
Experiment and Data Processing

A.1 Zero-Field Measurements

Experimental setups. In order to measure the complex conductivity of topological semimetals, two optical spectrometers were unitized. For the low-frequency part of the spectrum, typically from 30 or 50 cm^{-1} to 1000 cm^{-1}, a Bruker IFS 113v Fourier-transform infrared spectrometer was employed. At higher frequencies (usually, from 700 to 22 000 cm^{-1}, sometimes to 25 000 cm^{-1}) a Bruker Hyperion microscope attached to a Bruker Vertex 80v spectrometer was used. For all samples, we measured reflection from either as-grown or polished surfaces, as indicated in the relevant chapters. In order to obtain the absolute values of reflectivity R, we utilized either an in-situ gold overfilling technique [192] (for the reference measurements with the Bruker IFS 113v spectrometer) or freshly evaporated gold or coated silver mirrors (for the microscope measurements). The spectrometers were equipped with optical cryostats, allowing temperature-dependent measurements down to $T = 4.2$ K. The typical temperature range was 10–300 K. Additionally, room-temperature ellipsometry measurements up to 48 000 cm^{-1} were performed on some of the samples using a Woollam variable-angle spectroscopic ellipsometer with SiO_2 on Si substrate as reference. The magneto-optical measurements were made with the Bruker IFS 113v spectrometer at a fixed T and without any gold evaporation. Instead, the zero-field spectrum was used as a reference. Overall, the optical measurements were performed in a common way [75, 307]. The optical studies were accompanied by transport measurements (temperature-dependent longitudinal and Hall conductivity), for which home-made setups operating down to 2 K were utilized. The details of the optical and transport setup settings for each investigated compound are provided in the corresponding chapters below.

Data processing. The complex conductivity was obtained from the measured reflectivity via Kramers-Kronig analysis [75, 308]. Extrapolations to zero frequency were typically made using the Hagen-Rubens relation in accordance with

A. V. Pronin, *Linear Electrodynamic Response of Topological Semimetals*,
Springer Series in Solid-State Sciences 199,
https://doi.org/10.1007/978-3-031-35637-7

the temperature-dependent direct-current (dc) resistivity measurements. For some of the samples, a more elaborative extrapolation procedure with a preliminary Drude-Lorentz fit of the reflectivity spectra was unitized [146], see the relevant chapters below for details. The temperature-independent high-frequency extrapolations of the measured spectra were made by applying the procedure proposed by Tanner [145]. It utilizes the x-ray atomic scattering functions for calculating the optical response for frequencies from 21 to 30 000 eV and the free-electron $R \propto \omega^{-4}$ extrapolation for frequencies above 30 keV. In order to bridge the gap between our measurements and 21 eV, we used either published data obtained in ultra-high vacuum (if available) or interpolations suggested in [145].

A.2 Measurements in Magnetic Field

The majority of the magneto-optical measurements discussed in this book were performed in the far-infrared range (~30–700 cm^{-1}) using a commercial optical magnet (Spectromag from Oxford Instruments) with a superconducting split-coil magnet enabling magnetic fields of up to 7 T [163]. The coils are immersed in liquid He. The magnet has optical access and is coupled to a Bruker IFS 113v Fourier-transform spectrometer in such a way that the influence of the magnet stray field on the spectrometer optical elements is fully eliminated. Measurements with Faraday and both Voigt geometries are possible and the sample temperature can be varied down to 4.2. For the goals of this book, Voigt geometries were utilized and the sample temperatures were typically kept below 20 K, as indicated in the experimental Chapters. The ratios of the reflectivity in applied magnetic fields to the zero-field reflectivity were recorded. Additionally, optical measurements of TaAs at 20 T were performed using a combination of a resistive Bitter magnet and a far-infrared (21 cm^{-1}, 2.6 meV) free-electron laser as a light source, see Sect. 3.5.1.1 and [241] for details.

References

1. S. Murakami, Phase transition between the quantum spin Hall and insulator phases in 3D: emergence of a topological gapless phase. New J. Phys. **9**, 356 (2007)
2. A. Kobayashi et al., Massless fermions in organic conductor. J. Phys. Soc. Jpn. **76**, 034711 (2007)
3. X. Wan et al., Topological semimetal and Fermi-arc surface states in the electronic structure of pyrochlore iridates. Phys. Rev. B **83**, 205101 (2011)
4. A.A. Burkov, L. Balents, Weyl semimetal in a topological insulator multilayer. Phys. Rev. Lett. **107**, 127205 (2011)
5. A.A. Burkov, M.D. Hook, L. Balents, Topological nodal semimetals. Phys. Rev. B **84**, 235126 (2011)
6. J.L. Mañes, Existence of bulk chiral fermions and crystal symmetry. Phys. Rev. B **85**, 155118 (2012)
7. Z. Wang et al., Dirac semimetal and topological phase transitions in A_3Bi (A=Na, K, Rb). Phys. Rev. B **85**, 195320 (2012)
8. Z. Wang et al., Three-dimensional Dirac semimetal and quantum transport in Cd_3As_2. Phys. Rev. B **88**, 125427 (2013)
9. A.A. Soluyanov et al., Type-II Weyl semimetals. Nature **527**, 495 (2015)
10. T. Bzdušek et al., Nodal-chain metals. Nature **538**, 75 (2016)
11. B. Bradlyn et al., Beyond Dirac and Weyl fermions: unconventional quasiparticles in conventional crystals. Science **353**, aaf5037 (2016)
12. Z.K. Liu et al., A stable three-dimensional topological Dirac semimetal Cd_3As_2. Nat. Mater. **13**, 677 (2014)
13. S. Borisenko et al., Experimental realization of a three-dimensional Dirac semimetal. Phys. Rev. Lett. **113**, 027603 (2014)
14. M. Neupane et al., Observation of a three-dimensional topological Dirac semimetal phase in high-mobility Cd_3As_2. Nat. Commun. **5**, 3786 (2014)
15. B.Q. Lv et al., Experimental discovery of Weyl semimetal TaAs. Phys. Rev. X **5**, 031013 (2015)
16. S.-M. Huang et al., A Weyl fermion semimetal with surface Fermi arcs in the transition metal monopnictide TaAs class. Nat. Commun. **6**, 7373 (2015)
17. C. Shekhar et al., Extremely large magnetoresistance and ultrahigh mobility in the topological Weyl semimetal candidate NbP. Nat. Phys. **11**, 645 (2015)

A. V. Pronin, *Linear Electrodynamic Response of Topological Semimetals*, Springer Series in Solid-State Sciences 199,
https://doi.org/10.1007/978-3-031-35637-7

18. H. Weng et al., Weyl semimetal phase in noncentrosymmetric transition-metal monophosphides. Phys. Rev. X **5**, 011029 (2015)
19. S.-Y. Xu et al., Discovery of a Weyl fermion semimetal and topological Fermi arcs. Science **349**, 613 (2015)
20. D.-F. Xu et al., Observation of Fermi arcs in non-centrosymmetric Weyl semi-metal candidate NbP. Chin. Phys. Lett. **32**, 107101 (2015)
21. Z. Wang et al., Helicity-protected ultrahigh mobility Weyl fermions in NbP. Phys. Rev. B **93**, 121112 (2016)
22. L.M. Schoop et al., Dirac cone protected by non-symmorphic symmetry and three-dimensional Dirac line node in ZrSiS. Nat. Commun. **7**, 11696 (2016)
23. N.P. Armitage, E.J. Mele, A. Vishwanath, Weyl and Dirac semimetals in three-dimensional solids. Rev. Mod. Phys. **90**, 015001 (2018)
24. B.Q. Lv, T. Qian, H. Ding, Experimental perspective on three-dimensional topological semimetals. Rev. Mod. Phys. **93**, 025002 (2021)
25. Z. Zhu et al., Triple point topological metals. Phys. Rev. X **6**, 031003 (2016)
26. A.A. Burkov, Weyl metals. Annu. Rev. Condens. Matter Phys. **9**, 359 (2018)
27. W. Wu et al., Nodal surface semimetals: theory and material realization. Phys. Rev. B **97**, 115125 (2018)
28. J.-Z. Ma et al., Observation of a singular Weyl point surrounded by charged nodal walls in PtGa. Nat. Commun. **12**, 3994 (2021)
29. Z.-M. Yu et al., Encyclopedia of emergent particles in three-dimensional crystals. Sci. Bull. **67**, 375 (2022)
30. C.L. Kane, E.J. Mele, Z_2 topological order and the quantum spin Hall effect. Phys. Rev. Lett. **95**, 146802 (2005)
31. C.L. Kane, E.J. Mele, Quantum spin Hall effect in graphene. Phys. Rev. Lett. **95**, 226801 (2005)
32. B.A. Bernevig, S.-C. Zhang, Quantum spin Hall effect. Phys. Rev. Lett. **96**, 106802 (2006)
33. B.A. Bernevig, T.L. Hughes, S.-C. Zhang, Quantum spin Hall effect and topological phase transition in HgTe quantum wells. Science **314**, 1757 (2006)
34. L. Fu, C.L. Kane, Topological insulators with inversion symmetry. Phys. Rev. B **76**, 045302 (2007)
35. M. König et al., Quantum spin Hall insulator state in HgTe quantum wells. Science **318**, 766 (2007)
36. D. Hsieh et al., A topological Dirac insulator in a quantum spin Hall phase. Nature **452**, 970 (2008)
37. Y. Xia et al., Observation of a large-gap topological-insulator class with a single Dirac cone on the surface. Nat. Phys. **5**, 398 (2009)
38. H. Zhang et al., Topological insulators in Bi_2Se_3, Bi_2Te_3 and Sb_2Te_3 with a single Dirac cone on the surface. Nat. Phys. **5**, 438 (2009)
39. J.E. Moore, The birth of topological insulators. Nature **464**, 194 (2010)
40. T.O. Wehling, A.M. Black-Schaffer, A.V. Balatsky, Dirac materials. Adv. Phys. **63**, 1 (2014)
41. E.V. Gorbar et al., *Electronic Properties of Dirac and Weyl Semimetals*. (World Scientific, 2021)
42. C. Guo et al., Light control with Weyl semimetals. eLight **3**, 2 (2023)
43. D. Vanderbilt, *Berry Phases in Electronic Structure Theory*. (Cambridge University Press, 2018)
44. T. Frankel, *The Geometry of Physics: An Introduction*. (Cambridge University Press, 2011)
45. M. Nakahara, *Geometry, Topology and Physics*. (Institute of Physics Publishing, 2003)
46. B.A. Bernevig, T.L. Hughes, *Topological Insulators and Topological Super-Conductors*. (Princeton University Press, 2013)
47. R. Moessner, J.E. Moore, *Topological Phases of Matter*. (Cambridge University Press, 2021)
48. M.Z. Hasan, C.L. Kane, Topological insulators. Rev. Mod. Phys. **82**, 3045 (2010)
49. X.-L. Qi, S.-C. Zhang, Topological insulators and superconductors. Rev. Mod. Phys. **83**, 1057 (2011)

50. O. Vafek, A. Vishwanath, From high-Tc cuprates and graphene to topological insulators and Weyl semimetals. Annu. Rev. Condens. Matter Phys. **5**, 83 (2014)
51. C. Fang et al., Topological nodal line semimetals. Chin. Phys. B **25**, 117106 (2016)
52. F.D.M. Haldane, S. Raghu, Possible realization of directional optical waveguides in photonic crystals with broken time-reversal symmetry. Phys. Rev. Lett. **100**, 013904 (2008)
53. H. Wang et al., Topological photonic crystals: a review. Front. Optoelectron. **13**, 50 (2020)
54. P. McClarty, Topological magnons: a review. Annu. Rev. Condens. Matter Phys. **13**, 171 (2022)
55. X. Zhang et al., Topological sound. Commun. Phys. **1**, 97 (2018)
56. N.A. Spaldin, A beginner's guide to the modern theory of polarization. J. Solid State Chem. **195**, 2 (2012)
57. C. Nayak et al., Non-Abelian anyons and topological quantum computation. Rev. Mod. Phys. **80**, 1083 (2008)
58. A. Bohm et al., *The Geometric Phase in Quantum Systems*. (Springer, 2003)
59. Y.D. Chong, Berrys phase and the anomalous velocity of Bloch wavepackets. Phys. Rev. B **81**, 052303 (2010)
60. J.M. Ziman, *Principles of the Theory of Solids*. (Cambridge University Press, 1972)
61. N. Nagaosa et al., Anomalous Hall effect. Rev. Mod. Phys. **82**, 1539 (2010)
62. C.-Z. Chang et al., Experimental observation of the quantum anomalous Hall effect in a magnetic topological insulator. Science **340**, 167 (2013)
63. C.L. Kane, *Topological Band Theory and the* Z_2 *Invariant*. Chapter in Contemporary Concepts of Condensed Matter Science, vol. 6, Topological Insulators, ed. by M. Franz, L. Molenkamp (Elsevier, 2013)
64. Z.F. Ezawa, *Quantum Hall Effects: Recent Theoretical and Experimental Developments*. (World Scientific, 2013)
65. S.G. Davison, M. Stęślicka, *Basic Theory of Surface States*. (Clarendon Press, 1992)
66. B. Bradlyn et al., Topological quantum chemistry. Nature **547**, 298 (2017)
67. M.G. Vergniory et al., A complete catalogue of high-quality topological materials. Nature **566**, 480 (2019)
68. M.G. Vergniory et al., All topological bands of all nonmagnetic stoichiometric materials. Science **376**, 6595 (2022)
69. B.A. Bernevig et al., https://www.topologicalquantumchemistry.com
70. S.L. Adler, Axial-vector vertex in spinor electrodynamics. Phys. Rev. **177**, 2426 (1969)
71. J.S. Bell, R. Jackiw, A PCAC puzzle: $\pi^0 \to \gamma\gamma$ in the σ-model. Nuovo Cimento A **60**, 47 (1969)
72. H.B. Nielsen, M. Ninomiya, A no-go theorem for regularizing chiral fermions. Phys. Lett. B **105**, 219 (1981)
73. L.D. Landau, E.M. Lifshitz, *Electrodynamics of Continuous Media*. (Pergamon Press, 1984)
74. V.M. Galitskii, V.M. Ermachenko, *Macroscopic Electrodynamics [in Russian]*. (Vysshaya Shkola, 1988)
75. M. Dressel, G. Grüner, *Electrodynamics of Solids*. (Cambridge University Press, 2002)
76. A.K. Zvezdin, V.A. Kotov, *Modern Magnetooptics and Magnetooptical Materials*. (CRC Press, 2020)
77. D. Ivannikov et al., High-field ESR spectroscopy of the spin dynamics in $La_{1-x}Sr_xMnO_3$ ($x \leq 0.175$). Phys. Rev. B **65**, 214422 (2002)
78. A. Pimenov et al., Possible evidence for electromagnons in multiferroic manganites. Nat. Phys. **2**, 97 (2006)
79. P.Y. Yu, M. Cardona, *Fundamentals of Semiconductors: Physics and Materials Properties*. (Springer, 2010)
80. P.E.C. Ashby, J.P. Carbotte, Chiral anomaly and optical absorption in Weyl semimetals. Phys. Rev. B **89**, 245121 (2014)
81. J.P. Carbotte, Dirac cone tilt on interband optical background of type-I and type-II Weyl semimetals. Phys. Rev. B **94**, 165111 (2016)

82. C.J. Tabert et al., Optical and transport properties in three-dimensional Dirac and Weyl semimetals. Phys. Rev. B **93**, 085426 (2016)
83. C.J. Tabert, J.P. Carbotte, Optical conductivity of Weyl semimetals and signatures of the gapped semimetal phase transition. Phys. Rev. B **93**, 085442 (2016)
84. J.P. Carbotte, Optical response of a line node semimetal. J. Phys.: Condens. Matter **29**, 045301 (2017)
85. S.P. Mukherjee, J.P. Carbotte, Transport and optics at the node in a nodal loop semimetal. Phys. Rev. B **95**, 214203 (2017)
86. S.P. Mukherjee, J.P. Carbotte, Optical response in Weyl semimetal in model with gapped Dirac phase. J. Phys.: Condens. Matter **29**, 425301 (2017)
87. S.P. Mukherjee, J.P. Carbotte, Doping and tilting on optics in noncentrosymmetric multi-Weyl semimetals. Phys. Rev. B **97**, 045150 (2018)
88. J.P. Carbotte et al., Optical properties of a semi-Dirac material. Phys. Rev. B **99**, 115406 (2019)
89. J.P. Carbotte, E.J. Nicol, Signatures of merging Dirac points in optics and transport. Phys. Rev. B **100**, 035441 (2019)
90. P. Hosur, S.A. Parameswaran, A. Vishwanath, Charge transport in Weyl semimetals. Phys. Rev. Lett. **108**, 046602 (2012)
91. A. Bácsi, A. Virosztek, Low-frequency optical conductivity in graphene and in other scale-invariant two-band systems. Phys. Rev. B **87**, 125425 (2013)
92. Y. Xu, F. Zhang, C. Zhang, Structured Weyl points in spin-orbit coupled fermionic superfluids. Phys. Rev. Lett. **115**, 265304 (2015)
93. G. Xu et al., Chern semimetal and the quantized anomalous Hall effect in $HgCr_2Se_4$. Phys. Rev. Lett. **107**, 186806 (2011)
94. C. Fang et al., Multi-Weyl topological semimetals stabilized by point group symmetry. Phys. Rev. Lett. **108**, 266802 (2012)
95. S.-M. Huang et al., New type of Weyl semimetal with quadratic double Weyl fermions. PNAS **113**, 1180 (2016)
96. B. Singh et al., Tunable double-Weyl fermion semimetal state in the $SrSi_2$ materials class. Sci. Rep. **8**, 10540 (2018)
97. M.Á. Sánchez-Martínez, F. de Juan, A.G. Grushin, Linear optical conductivity of chiral multifold fermions. Phys. Rev. B **99**, 155145 (2019)
98. S. Ahn, E.J. Mele, H. Min, Optical conductivity of multi-Weyl semimetals. Phys. Rev. B **95**, 161112 (2017)
99. T. Ando, Y. Zheng, H. Suzuura, Dynamical conductivity and zero-mode anomaly in honeycomb lattices. J. Phys. Soc. Jpn. **71**, 1318 (2002)
100. K.F. Mak et al., Measurement of the optical conductivity of graphene. Phys. Rev. Lett. **101**, 196405 (2008)
101. A.B. Kuzmenko et al., Universal optical conductance of graphite. Phys. Rev. Lett. **100**, 117401 (2008)
102. S. Ahn, E.J. Mele, H. Min, Electrodynamics on Fermi cyclides in nodal line semimetals. Phys. Rev. Lett. **119**, 147402 (2017)
103. Y. Shao et al., Optical signatures of Dirac nodal lines in $NbAs_2$. PNAS **116**, 1168 (2019)
104. X. Huang et al., Observation of the chiral-anomaly-induced negative magnetoresistance in 3D Weyl semimetal TaAs. Phys. Rev. X **5**, 031023 (2015)
105. J. Xiong et al., Evidence for the chiral anomaly in the Dirac semimetal Na_3Bi. Science **350**, 413 (2015)
106. N. Kumar et al., Planar Hall effect in the Weyl semimetal GdPtBi. Phys. Rev. B **98**, 041103 (2018)
107. R.D. dos Reis et al., On the search for the chiral anomaly in Weyl semimetals: the negative longitudinal magnetoresistance. New J. Phys. **18**, 085006 (2016)
108. F. Arnold et al., Negative magnetoresistance without well-defined chirality in the Weyl semimetal TaP. Nat. Commun. **7**, 11615 (2016)

109. S. Liang et al., Experimental tests of the chiral anomaly magnetoresistance in the Dirac-Weyl semimetals Na_3Bi and GdPtBi. Phys. Rev. X **8**, 031002 (2018)
110. Q. Ma et al., Direct optical detection of Weyl fermion chirality in a topological semimetal. Nat. Phys. **13**, 842 (2017)
111. A.L. Levy et al., Optical evidence of the chiral magnetic anomaly in Weyl semimetal TaAs. Phys. Rev. B **101**, 125102 (2020)
112. X. Yuan et al., The discovery of dynamic chiral anomaly in a Weyl semimetal NbAs. Nat. Commun. **11**, 1259 (2020)
113. P. Hosur, X.-L. Qi, Tunable circular dichroism due to the chiral anomaly in Weyl semimetals. Phys. Rev. B **91**, 081106 (2015)
114. D.T. Son, B.Z. Spivak, Chiral anomaly and classical negative magnetoresistance of Weyl metals. Phys. Rev. B **88**, 104412 (2013)
115. E.D. Palik, J.K. Furdyna, Infrared and microwave magnetoplasma effects in semiconductors. Rep. Prog. Phys. **33**, 1193 (1970)
116. Y. Kim et al., Dirac line nodes in inversion-symmetric crystals. Phys. Rev. Lett. **115**, 036806 (2015)
117. S.T. Ramamurthy, T.L. Hughes, Quasitopological electromagnetic response of line-node semimetals. Phys. Rev. B **95**, 075138 (2017)
118. C. Fang et al., Topological nodal line semimetals with and without spin-orbital coupling. Phys. Rev. B **92**, 081201 (2015)
119. G. Bian et al., Topological nodal-line fermions in spin-orbit metal $PbTaSe_2$. Nat. Commun. **7**, 10556 (2016)
120. Y. Wu et al., Dirac node arcs in $PtSn_4$. Nat. Phys. **12**, 667 (2016)
121. Y. Chen, Y.-M. Lu, H.-Y. Kee, Topological crystalline metal in orthorhombic perovskite iridates. Nat. Commun. **6**, 6593 (2015)
122. R. Yu et al., Topological node-line semimetal and Dirac semimetal state in antiperovskite Cu_3PdN. Phys. Rev. Lett. **115**, 036807 (2015)
123. Y. Okamoto et al., Low carrier density metal realized in candidate line-node Dirac semimetals CaAgP and CaAgAs. J. Phys. Soc. Jpn. **85**, 123701 (2016)
124. L.S. Xie et al., A new form of $Ca_3P_2$with a ring of Dirac nodes. APL Mater. **3**, 083602 (2015)
125. Q. Xu et al., Two-dimensional oxide topological insulator with iron-pnictide superconductor LiFeAs structure. Phys. Rev. B **92**, 205310 (2015)
126. M. Neupane et al., Observation of topological nodal fermion semimetal phase in ZrSiS. Phys. Rev. B **93**, 201104 (2016)
127. C. Chen et al., Dirac line nodes and effect of spin-orbit coupling in the nonsymmorphic critical semimetals MSiS(M = Hf, Zr). Phys. Rev. B **95**, 125126 (2017)
128. A. Topp et al., Non-symmorphic band degeneracy at the Fermi level in ZrSiTe. New J. Phys. **18**, 125014 (2016)
129. M.-X. Yang, W. Luo, W. Chen, Quantum transport in topological nodal-line semimetals. Adv. Phys. X **7**(1), 2065216 (2022)
130. W. Chen, H.-Z. Lu, J.-M. Hou, Topological semimetals with a double-helix nodal link. Phys. Rev. B **96**, 041102 (2017)
131. Z. Yan et al., Nodal-link semimetals. Phys. Rev. B **96**, 041103 (2017)
132. M. Ezawa, Topological semimetals carrying arbitrary Hopf numbers: Fermi surface topologies of a Hopf link, Solomon's knot, trefoil knot, and other linked nodal varieties. Phys. Rev. B **96**, 041202 (2017)
133. D. Takane et al., Dirac-node arc in the topological line-node semimetal HfSiS. Phys. Rev. B **94**, 121108 (2016)
134. J. Hu et al., Evidence of topological nodal-line fermions in ZrSiSe and ZrSiTe. Phys. Rev. Lett. **117**, 016602 (2016)
135. L.M. Schoop et al., Tunable Weyl and Dirac states in the nonsymmorphic compound CeSbTe. Sci. Adv. **4**, eaar2317 (2018)
136. X. Wang et al., Evidence of both surface and bulk Dirac bands and anisotropic nonsaturating magnetoresistance in ZrSiS. Adv. Electron. Mater. **2**, 1600228 (2016)

137. R. Singha et al., Large nonsaturating magnetoresistance and signature of nondegenerate Dirac nodes in ZrSiS. PNAS **114**, 2468 (2017)
138. R. Sankar et al., Crystal growth of Dirac semimetal ZrSiS with high magnetoresistance and mobility. Sci. Rep. **7**, 40603 (2017)
139. M.N. Ali et al., Butterfly magnetoresistance, quasi-2D Dirac Fermi surface and topological phase transition in ZrSiS. Sci. Adv. **2**, e1601742 (2016)
140. S. Pezzini et al., Unconventional mass enhancement around the Dirac nodal loop in ZrSiS. Nat. Phys. **14**, 178 (2018)
141. M. Matusiak, J.R. Cooper, D. Kaczorowski, Thermoelectric quantum oscillations in ZrSiS. Nat. Commun. **8**, 15219 (2017)
142. J. Hu et al., Nearly massless Dirac fermions and strong Zeeman splitting in the nodal-line semimetal ZrSiS probed by de Haas-van Alphen quantum oscillations. Phys. Rev. B **96**, 045127 (2017)
143. M.B. Schilling et al., Flat optical conductivity in ZrSiS due to two-dimensional Dirac bands. Phys. Rev. Lett. **119**, 187401 (2017)
144. E. Uykur et al., Magneto-optical probe of the fully gapped Dirac band in ZrSiS. Phys. Rev. Res. **1**, 032015 (2019)
145. D.B. Tanner, Use of x-ray scattering functions in Kramers-Kronig analysis of reflectance. Phys. Rev. B **91**, 035123 (2015)
146. M.B. Schilling et al., Two-channel conduction in YbPtBi. Phys. Rev. B **95**, 155201 (2017)
147. J. Ebad-Allah et al., Chemical pressure effect on the optical conductivity of the nodal-line semimetals ZrSi$Y(Y$ = S, Se, Te) and ZrGe$Y(Y$ = S, Te). Phys. Rev. B **99**, 125154 (2019)
148. T. Habe, M. Koshino, Dynamical conductivity in the topological nodal- line semimetal ZrSiS. Phys. Rev. B **98**, 125201 (2018)
149. V.P. Gusynin, S.G. Sharapov, J.P. Carbotte, ac conductivity of graphene: from tight-binding model to 2+1-dimensional quantum electrodynamics. Int. J. Mod. Phys. B **21**, 4611 (2007)
150. O.V. Kotov, Y.E. Lozovik, Dielectric response and novel electromagnetic modes in three-dimensional Dirac semimetal films. Phys. Rev. B **93**, 235417 (2016)
151. G.S. Jenkins et al., Three-dimensional Dirac cone carrier dynamics in Na_3Bi and Cd_3As_2. Phys. Rev. B **94**, 085121 (2016)
152. W. Zhou et al., Lattice dynamics of Dirac node-line semimetal ZrSiS. Phys. Rev. B **96**, 064103 (2017)
153. A. Topp et al., Surface floating 2D bands in layered nonsymmorphic semimetals: ZrSiS and related compounds. Phys. Rev. X **7**, 041073 (2017)
154. C. Klingshirn. *Semiconductor Optics*. (Springer, 2007)
155. E.I. Rashba, Giant oscillator strengths associated with exciton complexes (review). Sov. Phys. Semicond. **8**, 807 (1975)
156. D.A. Bandurin et al., Negative local resistance caused by viscous electron backflow in graphene. Science **351**, 1055 (2016)
157. J. Crossno et al., Observation of the Dirac fluid and the breakdown of the Wiedemann-Franz law in graphene. Science **351**, 1058 (2016)
158. J. Gooth et al., Thermal and electrical signatures of a hydrodynamic electron fluid in tungsten diphosphide. Nat. Commun. **9**, 4093 (2018)
159. M. Orlita et al., Carrier scattering from dynamical magnetoconductivity in quasineutral epitaxial graphene. Phys. Rev. Lett. **107**, 216603 (2011)
160. M. Orlita et al., Classical to quantum crossover of the cyclotron resonance in graphene: a study of the strength of intraband absorption. New J. Phys. **14**, 095008 (2012)
161. Z.-G. Chen et al., Spectroscopic evidence for bulk-band inversion and three-dimensional massive Dirac fermions in $ZrTe_5$. PNAS **114**, 816 (2017)
162. Z.-G. Chen et al., Observation of an intrinsic bandgap and Landau level renormalization in graphene/boron-nitride heterostructures. Nat. Commun. **5**, 4461 (2014)
163. D.J. Neubauer, Optical and magneto-optical investigations on 3D Dirac and Weyl semimetals. Ph.D. Thesis, Universität Stuttgart, Germany, 2017. https://doi.org/10.18419/opus-9166

164. X. Chen et al., Dirac edges of fractal magnetic minibands in graphene with hexagonal Moiré superlattices. Phys. Rev. B **89**, 075401 (2014)
165. M.S. Lodge et al., Observation of effective pseudospin scattering in ZrSiS. Nano Lett. **17**(12), 7213 (2017)
166. I.O. Nedoliuk et al., Colossal infrared and terahertz magneto-optical activity in a two-dimensional Dirac material. Nat. Nanotechnol. **14**, 756 (2019)
167. W. Duan et al., Accurate magneto-optical determination of radius of topological nodal-ring semimetals. Phys. Rev. B **99**, 045124 (2019)
168. M. Orlita et al., Observation of three-dimensional massless Kane fermions in a zinc-blende crystal. Nat. Phys. **10**, 233 (2014)
169. T. Timusk et al., Three-dimensional Dirac fermions in quasicrystals as seen via optical conductivity. Phys. Rev. B **87**, 235121 (2013)
170. R.Y. Chen et al., Optical spectroscopy study of the three-dimensional Dirac semimetal $ZrTe_5$. Phys. Rev. B **92**, 075107 (2015)
171. Q. Li et al., Chiral magnetic effect in $ZrTe_5$. Nat. Phys. **12**, 550 (2016)
172. D. Neubauer et al., Interband optical conductivity of the [001]-oriented Dirac semimetal Cd_3As_2. Phys. Rev. B **93**, 121202 (2016)
173. R. Kemmler et al., Free-carrier dynamics in Au_2Pb probed by optical conductivity measurements. J. Phys. Condens. Matter **30**, 485403 (2018)
174. D. Neubauer et al., Optical conductivity of the Weyl semimetal NbP. Phys. Rev. B **98**, 195203 (2018)
175. S. Polatkan et al., Magneto-optics of a Weyl semimetal beyond the conical band approximation: case study of TaP. Phys. Rev. Lett. **124**, 176402 (2020)
176. A. Yaresko, A.V. Pronin, Low-energy optical conductivity of TaP: comparison of theory and experiment. Crystals **11**, 56 (2021)
177. S. Jeon et al., Landau quantization and quasiparticle interference in the three-dimensional Dirac semimetal Cd_3As_2. Nat. Mater. **13**, 851 (2014)
178. L.P. He et al., Quantum transport evidence for the three-dimensional Dirac semimetal phase in Cd_3As_2. Phys. Rev. Lett. **113**, 246402 (2014)
179. T. Liang et al., Ultrahigh mobility and giant magnetoresistance in the Dirac semimetal Cd_3As_2. Nat. Mater. **14**, 280 (2015)
180. M.N. Ali et al., The crystal and electronic structures of Cd_3As_2, the three-dimensional electronic analogue of graphene. Inorg. Chem. **53**, 4062 (2014)
181. W.J. Turner, A.S. Fischler, W.E. Reese, Physical properties of several II-V semiconductors. Phys. Rev. **121**, 759 (1961)
182. M. Iwami, M. Yoshida, K. Kawabe, Infrared optical absorption in Cd_3As_2. Jpn. J. Appl. Phys. **12**, 1276 (1973)
183. M.J. Gelten et al., Optical verification of the valence band structure of cadmium arsenide. Solid State Commun. **33**, 833 (1980)
184. M.J. Gelten, Optical properties of cadmium phosphide and cadmium arsenide. *Doctor in de Technische Wetenschappen* Thesis, Technische Hogeschool, Eindhoven, The Netherlands, 1985
185. O. Houde et al., The infrared spectrum of Cd_3As_2. Solid State Commun. **57**, 247 (1986)
186. *Non-Tetrahedrally Bonded Elements and Binary Compounds I, Landolt- Börnstein-Group III Condensed Matter.* (Springer, 1998)
187. M. Zivitz, J.R. Stevenson, Optical properties of the Cd_3As_2-Cd_3P_2 semiconductor alloy system. Phys. Rev. B **10**, 2457 (1974)
188. K. Karnicka-Moscicka, A. Kisiel, L. Zdanowicz, Fundamental reflectivity spectra of monocrystalline and polycrystalline bulk Cd_3As_2. Solid State Commun. **44**, 373 (1982)
189. A.I. Kozlov, V.V. Sobolev, A.F. Knjazev, Reflection spectra of two polymorphic modifications of cadmium arsenide. Semiconductors **39**, 285 (2005)
190. C.P. Weber et al., Transient reflectance of photoexcited Cd_3As_2. Appl. Phys. Lett. **106**, 231904 (2015)

191. A. Rambo, M.J. Aubin, The preparation and passive annealing of Cd_3As_2 platelets. Can. J. Phys. **57**, 2093 (1979)
192. C.C. Homes et al., Technique for measuring the reflectance of irregular, submillimeter-sized samples. Appl. Opt. **32**, 2976 (1993)
193. J. Cao et al., Landau level splitting in Cd_3As_2 under high magnetic fields. Nat. Commun. **6**, 7779 (2015)
194. J. Feng et al., Large linear magnetoresistance in Dirac semimetal Cd_3As_2 with Fermi surfaces close to the Dirac points. Phys. Rev. B **92**, 081306 (2015)
195. L. Benfatto, E. Cappelluti, Spectroscopic signatures of massless gap opening in graphene. Phys. Rev. B **78**, 115434 (2008)
196. T. Morimoto, N. Nagaosa, Weyl Mott Insulator. Sci. Rep. **6**, 19853 (2016)
197. L.M. Schoop et al., Dirac metal to topological metal transition at a structural phase change in Au_2Pb and prediction of Z2 topology for the superconductor. Phys. Rev. B **91**, 214517 (2015)
198. D. Hamilton et al., Some new superconducting compounds. J. Phys. Chem. Solids **26**, 665 (1965)
199. K.W. Chen et al., Temperature-pressure phase diagram of cubic Laves phase Au_2Pb. Phys. Rev. B **93**, 045118 (2016)
200. Y.J. Yu et al., Fully gapped superconducting state in Au_2Pb: a natural candidate for topological superconductor. EPL **116**, 67002 (2016)
201. Y. Xing et al., Superconductivity in topologically nontrivial material Au_2Pb. npj Quantum Mater **1**, 16005 (2016)
202. Y. Wu et al., Electronic structure of the topological superconductor candidate Au_2Pb. Phys. Rev. B **98**, 161107 (2018)
203. L. Fu, C.L. Kane, Superconducting proximity effect and Majorana fermions at the surface of a topological insulator. Phys. Rev. Lett. **100**, 096407 (2008)
204. D. Wu et al., Optical investigations of the normal and superconducting states reveal two electronic subsystems in iron pnictides. Phys. Rev. B **81**, 100512 (2010)
205. D. Neubauer et al., Optical properties of superconducting $EuFe_2(As_{1-x}P_x)_2$. Phys. Status Solidi B **254**, 1600148 (2017)
206. N.E. Hussey, K. Takenaka, H. Takagi, Universality of the Mott-Ioffe-Regel limit in metals. Philos. Mag. **84**, 2847 (2004)
207. D.N. Basov, T. Timusk, Electrodynamics of high-T_c superconductors. Rev. Mod. Phys. **77**, 721 (2005)
208. P. Kostic et al., Non-Fermi-liquid behavior of $SrRuO_3$: evidence from infrared conductivity. Phys. Rev. Lett. **81**, 2498 (1998)
209. K. Takenaka, Y. Sawaki, S. Sugai, Incoherent-to-coherent crossover of optical spectra in $La_{0.825}Sr_{0.175}MnO_3$: temperature-dependent reflectivity spectra measured on cleaved surfaces. Phys. Rev. B **60**, 13011 (1999)
210. R.P.S.M. Lobo et al., Optical conductivity evidence of clean-limit superconductivity in LiFeAs. Phys. Rev. B **91**, 174509 (2015)
211. A.F. Ioffe, A.R. Regel, Non-crystalline, amorphous and liquid electronic semiconductors. Prog. Semicond. **4**, 237 (1960)
212. V.J. Emery, S.A. Kivelson, Superconductivity in bad metals. Phys. Rev. Lett. **74**, 3253 (1995)
213. S. Das Sarma, E.H. Hwang, Collective modes of the massless Dirac plasma. Phys. Rev. Lett. **102**, 206412 (2009)
214. R. Sachdeva et al., Plasmon modes of a massive Dirac plasma, and their superlattices. Phys. Rev. B **91**, 205426 (2015)
215. Y. Sun, S.-C. Wu, B. Yan, Topological surface states and Fermi arcs of the noncentrosymmetric Weyl semimetals TaAs, TaP, NbAs, and NbP. Phys. Rev. B **92**, 115428 (2015)
216. C.-C. Lee et al., Fermi surface interconnectivity and topology in Weyl fermion semimetals TaAs, TaP, NbAs, and NbP. Phys. Rev. B **92**, 235104 (2015)
217. K.-H. Ahn, K.-W. Lee, W.E. Pickett, Spin-orbit interaction driven collective electron-hole excitations in a noncentrosymmetric nodal loop Weyl semimetal. Phys. Rev. B **92**, 115149 (2015)

218. J. Klotz et al., Quantum oscillations and the Fermi surface topology of the Weyl semimetal NbP. Phys. Rev. B **93**, 121105 (2016)
219. S.-C. Wu et al., Hidden type-II Weyl points in the Weyl semimetal NbP. Phys. Rev. B **96**, 165113 (2017)
220. D. Grassano et al., Validity of Weyl fermion picture for transition metals monopnictides TaAs, TaP, NbAs, and NbP from ab initio studies. Sci. Rep. **8**, 3534 (2018)
221. J. Martin, R. Gruehn, Synthesis of transition-metal phosphides from their oxides using the chemical-transport (CVT) method. Solid State Ionics **43**, 19 (1990)
222. C. Zhang et al., Large magnetoresistance over an extended temperature regime in monophosphides of tantalum and niobium. Phys. Rev. B **92**, 041203 (2015)
223. A.B. Kuzmenko, Kramers-Kronig constrained variational analysis of optical spectra. Rev. Sci. Instrum. **76**, 083108 (2005)
224. A.F. Santander-Syro et al., In-plane electrodynamics of the superconductivity in $Bi_2Sr_2CaCu_2O_{8+\delta}$: energy scales and spectral weight distribution. Phys. Rev. B **70**, 134504 (2004)
225. G. Chanda et al., Optical study of superconducting Pr_2CuO_x with $x \simeq 4$. Phys. Rev. B **90**, 024503 (2014)
226. J. Xu et al., Crystal structure, electrical transport, and magnetic properties of niobium monophosphide. Inorg. Chem. **35**, 845 (1996)
227. O.K. Andersen, Linear methods in band theory. Phys. Rev. B **12**, 3060 (1975)
228. V. Antonov, B. Harmon, A. Yaresko, *Electronic Structure and Magneto-Optical Properties of Solids*. (Kluwer Academic Publishers, 2004)
229. J.P. Perdew, Y. Wang, Accurate and simple analytic representation of the electron-gas correlation energy. Phys. Rev. B **45**, 13244 (1992)
230. P.E. Blöchl, O. Jepsen, O.K. Andersen, Improved tetrahedron method for Brillouin-zone integrations. Phys. Rev. B **49**, 16223 (1994)
231. D. Chaudhuri et al., Optical investigation of the strong spin-orbit-coupled magnetic semimetal $YbMnBi_2$. Phys. Rev. B **96**, 075151 (2017)
232. B. Xu et al., Observation of the chiral-anomaly-induced negative magnetoresistance in 3D Weyl semimetal TaAs. Nat. Commun. **8**, 14933 (2017)
233. D. Chang et al., Phonon and thermal expansion properties in Weyl semimetals MX (M = Nb, Ta; X = P, As): ab initio studies. Phys. Chem. Chem. Phys. **18**, 14503 (2016)
234. H.W. Liu et al., Comparative Raman study of Weyl semimetals TaAs, NbAs, TaP and NbP. J. Phys.: Condens. Matter **28**, 295401 (2016)
235. B. Xu et al., Optical spectroscopy of the Weyl semimetal TaAs. Phys. Rev. B **93**, 121110 (2016)
236. S.-I. Kimura et al., Optical signature of Weyl electronic structures in tantalum pnictides TaPn (Pn = P, As). Phys. Rev. B **96**, 075119 (2017)
237. A.J. Frenzel et al., Anisotropic electrodynamics of type-II Weyl semimetal candidate WTe_2. Phys. Rev. B **95**, 245140 (2017)
238. F. Arnold et al., Chiral Weyl pockets and Fermi surface topology of the Weyl semimetal TaAs. Phys. Rev. Lett. **117**, 146401 (2016)
239. J. Willerstrom, Stacking disorder in NbP, TaP, NbAs and TaAs. J. Less-Common Met. **99**, 273 (1984)
240. T. Besara et al., Coexistence of Weyl physics and planar defects in the semimetals TaP and TaAs. Phys. Rev. B **93**, 245152 (2016)
241. M. Ozerov et al., A THz spectrometer combining the free electron laser FLARE with 33 T magnetic fields. Appl. Phys. Lett. **110**, 094106 (2017)
242. F. Hütt et al., Terahertz transmission through TaAs single crystals in simultaneously applied magnetic and electric fields: possible optical signatures of the chiral anomaly in a Weyl semimetal. Results Phys. **15**, 102630 (2019)
243. C. Shekhar, V. Süß, M. Schmidt, Mobility induced unsaturated high linear magnetoresistance in transition-metal monopnictides Weyl semimetals (2016). arXiv:1606.06649
244. L. Wollmann et al., Heusler 4.0: tunable materials. Annu. Rev. Mater. Res. **47**, 247 (2017)

245. K. Manna et al., Heusler. Weyl and Berry. Nat. Rev. Mater. **3**, 244 (2018)
246. F. Casper et al., Half-Heusler compounds: novel materials for energy and spintronic applications. Semicond. Sci. Technol. **27**, 063001 (2012)
247. S. Chadov et al., Tunable multifunctional topological insulators in ternary Heusler compounds. Nat. Mater. **9**, 541 (2010)
248. H. Lin et al., Half-Heusler ternary compounds as new multifunctional experimental platforms for topological quantum phenomena. Nat. Mater. **9**, 546 (2010)
249. D. Xiao et al., Half-Heusler compounds as a new class of three-dimensional topological insulators. Phys. Rev. Lett. **105**, 096404 (2010)
250. W. Al-Sawai et al., Topological electronic structure in half-Heusler topological insulators. Phys. Rev. B **82**, 125208 (2010)
251. C. Liu et al., Metallic surface electronic state in half-Heusler compounds *R*PtBi (*R* = Lu, Dy, Gd). Phys. Rev. B **83**, 205133 (2011)
252. Z.K. Liu et al., Observation of unusual topological surface states in half- Heusler compounds *Ln*PtBi (*Ln* = Lu, Y). Nat. Commun. **7**, 12924 (2016)
253. J.A. Logan et al., Observation of a topologically non-trivial surface state in half-Heusler PtLuSb (001) thin films. Nat. Commun. **7**, 11993 (2016)
254. C. Shekhar et al., Anomalous Hall effect in Weyl semimetal half-Heusler compounds *R*PtBi (*R* = Gd and Nd). PNAS **115**, 9140 (2018)
255. M.M. Hirschberger et al., The chiral anomaly and thermopower of Weyl fermions in the half-Heusler GdPtBi. Nat. Mater. **15**, 1161 (2016)
256. H. Yang et al., Prediction of triple point fermions in simple half-Heusler topological insulators. Phys. Rev. Lett. **119**, 136401 (2017)
257. G.W. Winkler, S. Singh, A.A. Soluyanov, Topology of triple-point metals. Chin. Phys. B **28**, 077303 (2019)
258. F. Hütt et al., Linear-in-frequency optical conductivity in GdPtBi due to transitions near the triple points. Phys. Rev. Lett. **121**, 176601 (2018)
259. P.C. Canfield et al., Magnetism and heavy fermion-like behavior in the *R*BiPt series. J. Appl. Phys. **70**, 5800 (1991)
260. Z.V. Popović et al., Vibrational properties of *Me*NiSn (*Me* = Ti, Zr, Hf). Solid State Commun. **74**, 829 (1990)
261. P. Hermet, K. Niedziolka, P. Jund, A first-principles investigation of the thermodynamic and mechanical properties of NiTiSn Heusler and half- Heusler materials. RSC Adv. **3**, 22176 (2013)
262. C. Çoban, K. Çolakoğlu, and Y. Ö. Çiftçi, First principles study of the structural, mechanical, phonon, optical, and thermodynamic properties of half-Heusler (HH) compound NbFeSb. Phys. Scr. **90**, 095701 (2015)
263. D. Shrivastava, S.P. Sanyal, Electronic structure and superconducting behaviour of LuPtBi half-Heusler compound: a first principle study. Physica C **544**, 22 (2018)
264. V.N. Antonov et al., Calculation of the magneto-optical properties of ferromagnetic metals using the spin-polarized relativistic LMTO method. J. Magn. Magn. Mater. **146**, 205 (1995)
265. J.P. Perdew, K. Burke, M. Ernzerhof, Generalized gradient approximation made simple. Phys. Rev. Lett. **77**, 3865 (1996)
266. X. Wang, J. Callaway, Band structure of nickel: spin-orbit coupling, the Fermi surface, and the optical conductivity. Phys. Rev. B **9**, 4897 (1974)
267. Z. Fisk et al., Massive electron state in YbBiPt. Phys. Rev. Lett. **67**, 3310 (1991)
268. R. Movshovich et al., Fermi surface instability and symmetry breaking in heavy-fermion compound YbBiPt. Phys. Rev. Lett. **73**, 492 (1994)
269. P.M. Oppeneer et al., Fermi level pinning of a massive electron state in YbBiPt. Phys. Rev. Lett. **78**, 4079 (1997)
270. E.D. Mun et al., Magnetic-field-tuned quantum criticality of the heavy-fermion system YbPtBi. Phys. Rev. B **87**, 075120 (2013)
271. E. Mun et al., Quantum oscillations in the heavy-fermion compound YbPtBi. Phys. Rev. B **92**, 085135 (2015)

272. B.G. Ueland et al., Fragile antiferromagnetism in the heavy-fermion compound YbBiPt. Phys. Rev. B **89**, 180403 (2014)
273. B.G. Ueland et al., Magnetic-field effects on the fragile antiferromagnetism in YbBiPt. Phys. Rev. B **99**, 184431 (2019)
274. P.C. Canfield, Z. Fisk, Growth of single crystals from metallic fluxes. Philos. Mag. B **65**, 1117 (1992)
275. R.A. Robinson et al., Crystallography and magnetism of the heavy-fermion compound YbBiPt. Phys. Rev. B **50**, 9595 (1994)
276. B.G. Ueland et al., High-resolution x-ray diffraction study of the heavy- fermion compound YbBiPt. Phys. Rev. B **92**, 184111 (2015)
277. Z. Hou et al., Large low-field positive magnetoresistance in nonmagnetic half-Heusler ScPtBi single crystal. Appl. Phys. Lett. **107**, 202103 (2015)
278. O. Pavlosiuk et al., Antiferromagnetism and superconductivity in the half-Heusler semimetal HoPdBi. Sci. Rep. **6**, 18797 (2016)
279. G. Chang et al., Topological quantum properties of chiral crystals. Nat. Mater. **17**, 978 (2018)
280. G. Chang et al., Unconventional chiral fermions and large topological Fermi arcs in RhSi. Phys. Rev. Lett. **119**, 206401 (2017)
281. P. Tang, Q. Zhou, S.-C. Zhang, Multiple types of topological fermions in transition metal silicides. Phys. Rev. Lett. **119**, 206402 (2017)
282. D.S. Sanchez et al., Topological chiral crystals with helicoid-arc quantum states. Nature **567**, 500 (2019)
283. Z. Rao et al., Observation of unconventional chiral fermions with long Fermi arcs in CoSi. Nature **567**, 496 (2019)
284. N.B.M. Schröter et al., Chiral topological semimetal with multifold band crossings and long Fermi arcs. Nat. Phys. **15**, 759 (2019)
285. N.B.M. Schröter et al., Observation and control of maximal Chern numbers in a chiral topological semimetal. Science **369**, 179 (2020)
286. D. Takane et al., Observation of chiral fermions with a large topological charge and associated Fermi-arc surface states in CoSi. Phys. Rev. Lett. **122**, 076402 (2019)
287. M. Yao et al., Observation of giant spin-split Fermi-arc with maximal Chern number in the chiral topological semimetal PtGa. Phys. Rev. B **11**, 2033 (2020)
288. P. Sessi et al., Handedness-dependent quasiparticle interference in the two enantiomers of the topological chiral semimetal PdGa. Nat. Commun. **11**, 3507 (2020)
289. A.V. Pronin, M. Dressel, Nodal semimetals: a survey on optical conductivity. Phys. Status Solidi B **258**, 2000027 (2021)
290. L.Z. Maulana et al., Optical conductivity of multifold fermions: the case of RhSi. Phys. Rev. Res. **2**, 023018 (2020)
291. L.Z. Maulana et al., Broadband optical conductivity of the chiral multifold semimetal PdGa. Phys. Rev. B **103**, 115206 (2021)
292. F. de Juan et al., Quantized circular photogalvanic effect in Weyl semimetals. Nat. Commun. **8**, 15995 (2017)
293. D. Rees et al., Helicity-dependent photocurrents in the chiral Weyl semimetal RhSi. Sci. Adv. **6**, eaba0509 (2020)
294. Z. Li et al., Optical response of the chiral topological semimetal RhSi. Phys. Rev. B **100**, 155201 (2019)
295. Z. Qiu et al., Observation of a topological nodal-line semimetal in $YbMnSb_2$ through optical spectroscopy. Phys. Rev. B **100**, 125136 (2019)
296. T. Habe, Dynamical conductivity in the multiply degenerate point-nodal semimetal CoSi. Phys. Rev. B **100**, 245131 (2019)
297. A. Damascelli et al., Infrared spectroscopic study of phonons coupled to charge excitations in FeSi. Phys. Rev. B **55**, R4863 (1997)
298. P. Giannozzi et al., QUANTUM ESPRESSO: a modular and open-source software project for quantum simulations of materials. J. Phys. Condens. Matter **21**, 395502 (2009)

299. K. Schwarz, P. Blaha, Solid state calculations using WIEN2k. Comput. Mater. Sci. **28**, 259 (2003)
300. S. Polatkan, E. Uykur, Optical response of chiral multifold semimetal PdGa. Crystals **11**, 80 (2021)
301. D. van der Marel et al., Spin, charge, and bonding in transition metal mono-silicides. Phys. B **244**, 138 (1998)
302. B. Xu et al., Optical signatures of multifold fermions in the chiral topological semimetal CoSi. PNAS **117**, 27104 (2020)
303. Z. Ni et al., Linear and nonlinear optical responses in the chiral multifold semimetal RhSi. npj Quantum Mater. **5**, 96 (2020)
304. R. Yang et al., Spin-canting-induced band reconstruction in the Dirac material $Ca_{1-x}Na_xMnBi_2$. Phys. Rev. Lett. **124**, 137201 (2020)
305. C.C. Homes, T. Wolf, C. Meingast, Anisotropic optical properties of detwinned $BaFe_2As_2$. Phys. Rev. B **102**, 155135 (2020)
306. M. Corasaniti et al., Fermi surface gapping in the Dirac material $Ca_{1-x}Na_xMnBi_2$. Phys. Rev. B **100**, 041107 (2019)
307. D.H. Martin (ed.), *Spectroscopic Techniques for Far Infra-Red, Submillimetre and Millimetre Waves* (North-Holland Publishing, 1967)
308. F. Wooten. *Optical Properties of Solids*. (Academic Press, 1972)

MIX
Papier aus verantwortungsvollen Quellen
Paper from responsible sources
FSC® C105338

If you have any concerns about our products,
you can contact us on
ProductSafety@springernature.com

In case Publisher is established outside the EU,
the EU authorized representative is:
Springer Nature Customer Service Center GmbH
Europaplatz 3, 69115 Heidelberg, Germany

Printed by Libri Plureos GmbH
in Hamburg, Germany